Flight Safety in General Aviation

R. D. Campbell

Collins Professional Books

Collins Books
William Collins Sons & Co. Ltd
8 Grafton Street, London W1X 3LA

First published in Great Britain by
Collins Professional Books 1987

Distributed in the United States of America
by Sheridan House, Inc.

Copyright © R. D. Campbell 1987

British Library Cataloguing in Publication Data
Campbell, R.D.
Flight Safety in general aviation.
1. Aeronautics—Safety measures
I. Title
363.1'2475   TL553.5

ISBN 0–00–383306–2

Typeset by V & M Graphics Ltd, Aylesbury, Bucks
Printed and bound in Great Britain by
Mackays of Chatham, Kent

All rights reserved. No part of this publication may
be reproduced, stored in a retrieval system or transmitted,
in any form, or by any means, electronic, mechanical,
photocopying, recording or otherwise, without the prior
permission of the publishers.

Front cover photo by courtesy of the Mission Aviation
Fellowship, a world wide missionary aviation organisation
made up of 17 autonomous national bodies working on the
same principle and in close co-operation. MAF around the
globe are currently operating a fleet of 140 aircraft in 30
countries.
    MAF aeroplanes and helicopters provide life saving aerial
support to missionaries and church workers, including relief
and medical aid, and giving educational and agricultural
assistance to people in need.
    In 1986 the MAF in the United Kingdom were awarded the
prestigous Lennox Boyd Trophy by the Aircraft Owners and
Pilots Association (UK). The award was made in recognition of
the consistently high standard of training to prepare pilots for
operations under the most difficult circumstances from curved
and sloping airstrips situated in some of the world's most
inhospitable territories to aviation. During all these operations
the MAF has maintained a most excellent safety record.

Contents

Preface		vi
Acknowledgements		viii
1	How safe is general aviation?	1
	Accidents by types of licence held	2
	Accidents by phase of flight	2
	Accidents relative to classification of aviation activity	3
	The development of general aviation	10
2	Motor skills and human factors	14
	High-risk phases of flight	20
	The most common causal factors	22
	Conditions which aggravate pilot error	26
3	Preflight planning and preparation	30
	Flight planning considerations	32
	Pre-flight preparation is the foundation of safe flying	39
	Reducing the risk of inadequate preflight preparation	40
4	Take-off and landing – performance aspects	43
	Factors affecting take-off performance	44
	Factors affecting landing performance	51
	The size of the training aerodrome	52
	The problems of the short take-off technique	53
	The dangers of compulsive actions	55
	Reducing the risk of a performance accident	59
5	The take-off – procedures and control	61
	Use of checklists	61
	During the power checks	62
	Pre-take-off checks	65
	The use of flap during take-off	65
	Crosswind factors during take-off	68
	Wake turbulence	71
	Birdstrikes	79
	Reducing the risk during take-off	82

6	The approach and landing – procedures and control	84
	Aerodrome approach checks	85
	The precision approach	91
	Reducing the risk during the approach and landing	100
7	Fuel management	103
	Complacency	103
	Misunderstanding	104
	Lack of care	105
	Range and endurance	107
	Fuel quantity – the physical check	107
	Fuel indicating systems and inherent errors	110
	Fuel tank selection	111
	Fuel contamination	113
	Fuel tank venting	115
	Frequency of powerplant failures	117
	Reducing the risk of having a fuel-related accident	121
8	Weather considerations	123
	Height of the cloud base	126
	Visibility	127
	Thunderstorms and line squalls	129
	Airframe icing	129
	Weather problems start on the ground	130
	Reducing the risk of accidents due to weather	138
9	The influence of pilot distractions on aircraft accidents	143
	Causal factors	143
	Planning the cockpit workload	144
	Distractions leading to loss of control	145
	Reducing the risk of distraction-type accidents	146
10	Prevention is better than cure	147
	Aircraft maintenance	147
	The preflight inspection	148
	Carburettor icing accidents	153
	Formation of engine icing	156
	Recognition and prevention of engine icing	160
	Reducing the risk of engine icing	167
	The forced landing situation	168
	Engine failure – remedial actions	172
11	Stall/spin accidents	180
	The past record	180
	Spin recovery – training limitations	182

	Reducing the risk of a stall/spin accident	185
12	The aircraft may be fit to fly, but what about you?	187
	Fatigue	187
	Hypoxia	188
	Hyperventilation	189
	Alcohol	190
	Drugs	190
	Disorientation	190
	Carbon monoxide	191
	Vision	192
	Middle ear discomfort or pain	192
	Recognition of physical and mental fitness	193

In conclusion   194

Bibliography   195

Index   197

Preface

Through the years considerable research has been conducted and many books have been written with the object of improving flight safety. No doubt if all the knowledge and advice given in these works had been digested and acted upon by pilots and others engaged at the sharp end of aviation, the accident record would by now be greatly improved. Sadly this is not the case. A review of the accident statistics reveals that, whilst the record does show an improvement, during the last 20 years or so it has been on a very small scale.

The reasons for this are many, but in the end, it has to be appreciated that pilots, engineers, air traffic controllers and others who are close to the operational side of aviation cannot possibly absorb, understand, retain and recall all the available information, particularly when put forward from highly erudite sources. This comment is not a criticism of the excellent work being done in the field of aviation safety, nor of the past publications concerning this subject, it is merely an admission of the facts of life as seen and experienced by one who has been actively flying for 45 years and who in the last 25 years has been primarily engaged in the training and testing of pilots and instructors in civil aviation.

This book may be considered a departure from previous works concerning air safety, and it is aimed at the general aviation pilot. It does not go into analytical concepts of psychology, physiology, psychomotor processes, research perspectives, ergonomics and the like, but rather it tackles the subject of safety in more practical terms, in a way which I consider might be more suited to a pilot's thought processes, and operational habits, as they apply to today's aircraft and the environment in which flying activities are conducted.

When commencing to write a book about aviation safety, one becomes very conscious of the difficulties of putting actual flying experience and the lessons learned from it in such a way that the reader will neither become bored nor feel that his own flying ability is being adversely criticised. One is also aware of the fact that a book which is aimed at counselling pilots to fly more safely must be factual, yet hold the reader's interest and create sufficient impact for the comments being made to be given serious consideration.

Reflecting on these points, however, I can also recall that there have been a number of occasions when, if I had heeded the advice put forward by some of my contemporaries in aviation at that time, I would not be here to write these words today. This comment is made because we must acknowledge that pilots are human and thus possess the in-built frailties and failings of humans, which means that we are all open to making mistakes or failing to put the facts together in the right order.

It is also an acknowledged fact that most of us live in countries where we can freely express our opinions on most subjects. However, when these opinions can affect the safety of others, care must be taken to ensure that whenever they are publicly expressed they are based upon facts – and even then one cannot always get it right because of the variations in pilot experience, skill, the type of aircraft being flown, and the conditions in which a particular situation occurs.

Nevertheless there is one fact that cannot be denied and this is that we are all at risk from the moment we are born to the moment we die, whether lying in our beds, working in the home, walking in the park, driving a car or flying an aeroplane. Thus whatever we do has an element of risk, we cannot eliminate it, but what is important is that we have the ability to reduce it. This will apply regardless of our chosen activity or inactivity, because we have the capability to control our actions and to establish and abide by personal rules.

Finally, and bearing these comments in mind, this book is not aimed at telling people how they should physically operate the controls in order to carry out a certain manoeuvre or procedure, whether in an aeroplane, helicopter or other category of aircraft. Instead it assumes that the reader has already received this training and has qualified as a pilot, or is undergoing training from a competent instructor. Therefore, except on a few occasions, the contents are directed towards a pilot's thinking and decision-making processes, as well as other items which, due to the constraints of time or for other reasons, may not have been covered or emphasised sufficiently during training.

R.D.C.

Acknowledgements

Acknowledgements are gratefully made to the UK Civil Aviation Authority, the Accident Investigation Branch of the UK Department of Transport, the National Transportation Board and Federal Aviation Administration of the USA, the Australian Bureau of Air Safety Investigation, the AOPA Air Safety Foundation, and the many persons whose help and assistance in compiling the information and photographs on the following pages were invaluable to the objective of this publication.

1
How safe is general aviation?

The ability to fly an aircraft brings with it a powerful feeling of personal achievement. However, the real objective of any pilot should be to develop the more difficult ability which permits him to *operate the aircraft with safety*. Once a pilot achieves this level of competence he or she can be truly adjudged master of the aeroplane and the environment in which it operates. It is this mastery which brings with it the knowledge of true achievement and the ability to safeguard the well-being of the passengers and general public alike.

This book is all about reducing risks when we fly; it is aimed at increasing competence, thus enabling a pilot to obtain greater pleasure and satisfaction from the investment that has already been made, in terms of finance and physical and mental effort which are always necessary in obtaining a pilot's licence.

On the following pages the actual risk factors are discussed and shown in each phase of flight. The various factors and pilot activities which lead to errors within each phase are then analysed in a manner which highlights how lack of knowledge and incorrect actions and procedures create risk, after which guidance is given on how such risks can be reduced.

Having accepted that aviation, just like other pursuits, involves risk in general terms, the first step we should take is to identify the specific areas and degrees of risk and to do this we must turn to past accident figures. However, being human we must also accept the fact that it is unlikely that we can be 100% on our guard against risk during every second we spend in an aircraft; thus it would be sensible to separate the highest risk areas first, i.e. those which in the past have most commonly caused fatalities or serious injuries, and it is here that we come up against our first real problem. This is because in reviewing the accident statistics we find they are published in different ways and use varying formulas to the extent that merely reading through them will be of little help to anyone wishing to reduce the risks which occur when operating an aeroplane.

For example, in aviation the published safety statistics in the main revolve around the number of aircraft accidents per hours or passenger kilometres flown. An interpretation of 'aviation safety' based upon these figures is consequently rather impractical because the true measure of safety ought surely to be assessed by 'hours of exposure relative to numbers of fatalities or injuries incurred', rather than by the number of accidents per hours flown. It should be noted that safety at sea or on the roads is measured by the number of persons killed or injured, and road safety statistics tell us that some 50,000 people are killed on the roads of Europe every year and a further million and a half suffer serious injury. Such figures must surely be a more accurate measure in determining road safety than comparisons between the number of damaged vehicles per number of hours or miles driven.

Accidents by types of licence held

The weakness of the aviation statistical accident recording system which places the emphasis upon number of accidents versus hours flown or kilometres travelled by passengers is evidenced by the way the daily press and news media are led into misinterpretations of actual air safety. A typical example of this was seen in an article in the Sunday supplement of a leading UK newspaper, which commenced as follows:

> ... of the 220 aviation accidents which occurred in the UK during 1980, 209 were attributable to private pilots and only 10 to professional pilots...'

Presumably the article was written to inform the public that private pilots were far less safe than those employed in the commercial sector of aviation, and the article went on to say that the safety figures were culled from the official CAA publication, *Accidents to Aircraft on the British Register (1980)*. However, this document is mainly a statistical survey and gives little information to the reader who tries to use it to measure safety factors in relation to numbers of persons injured or killed, or indeed even to see how accident figures in aviation might be reduced.

For example, upon a detailed examination of this CAA publication it becomes clear, after sifting through the mass of information given in the accident briefs and tables, that at least 44 of the accidents occurred with a professional pilot at the controls, and a further 26 were incurred by student pilots. Twenty-one of the accident briefs did not contain any information relating to the type of licence held, so in terms of established fact it could be said that only 129 accidents occurred at the hands of private pilots. However, a further scrutiny of the published information reduced this figure to 125 occasions when a private pilot was at the controls. This figure is quite different from the 209 attributed to private pilots by the author of the article in question and in fact represents a 40% reduction in the number of accidents which he lays at the door of the private pilot sector. (Though, to be fair, this information is not easily interpreted from the way the tables in this CAA publication are constructed.)

Accidents by phase of flight

There are of course many other considerations to be taken into account when one attempts to make comparisons between the safety factor of different aviation sectors, and not least of these is the type of flying activity and the risk element per phase of flight. In this latter respect it should be borne in mind that the largest percentage of all aircraft accidents occurs during the take-off and initial climb and the approach and landing phases of a flight; these two phases are thus the highest risk sectors of any flight and this fact is very significant when comparisons are being made between the air transport sector and general aviation. For example, the average stage length of an aircraft operating in the commercial air transportation sector is nearly 2 hours whereas the average stage length of a light aircraft flight is about 30 minutes. Therefore by the very nature of light aircraft operations the private pilot is exposed to the highest risk area four times as often as his professional counterpart. Although this is far from being the whole story it might be argued on this fact alone that the private pilot is equally as safe in his flying activities as the professional is in his. After all, if the 44 accidents by professional pilots were multiplied by a factor of 3? ... But perhaps that would be falling into the same trap of misinterpreting statistics and the meaning of safety.

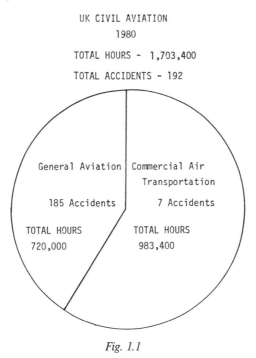

Fig. 1.1

Accidents relative to classification of aviation activity

The above example is given to point out that the value of statistics is lost if they are not interpreted correctly. To further illustrate this point, if we were to attempt to compare safety between the airline passenger services and general aviation* in the United Kingdom in relation to hours flown per accident, then by referring to Table 10 of the 1980 Annual Survey we would come up with the conclusions shown in Fig. 1.1.

Accidents to airline passenger services during 1980 numbered 7 for a total revenue hours flown of 983,400, which equates to one accident every 140,485 hours flown by fixed wing public transport aircraft over 2,300 kg. The same publication also determines that general aviation aircraft under 2,300 kg (including helicopters) had 185 accidents in 720,000 hours flown, and this equates to one accident for every 3,890 hours flown (Fig. 1.2).

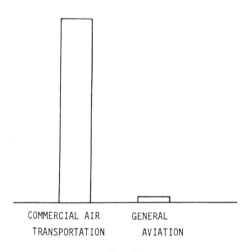

Fig. 1.2

The immediate impression gained from the above information is that the flying conducted within the public transport services is 36 times safer than that done by the general aviation sector. However, if we were to measure safety in relation to injury or death to persons rather than just counting up the number of occasions on which an aircraft is damaged, we would find upon examining the CAA's Annual Survey of Accidents that quite a different picture emerges.

Note: ICAO defines general aviation in Annex 6 Part 2 as '... all civil aviation operations other than scheduled air services and non-scheduled air transport operations for remuneration or hire'. This definition effectively covers all forms of private flying including those elements known as executive, training, club, and group activities.

For example, in 1980, although only 7 accidents took place to aircraft engaged in airline passenger services the number of people killed in these accidents was 156. If this figure is divided into the number of hours flown (983,400) this will equate to 1 person killed for every 6,304 hours flown.

By comparison general aviation, including both fixed wing and helicopters below 2,300 kg, flew 720,000 hours during which only 22 persons were killed and this equates to 1 person killed for every 32,727 hours flown.

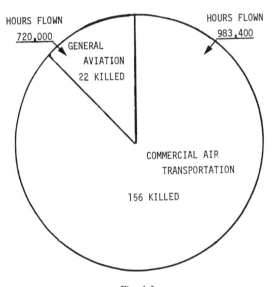

Fig. 1.3

As can be seen from Fig. 1.3 there were 7 times more people killed in the commercial air transport sector than in general aviation, yet the commercial air transport sector flew less than 1.5 times more hours than general aviation. So when we compare safety in human terms relative to hours flown we see that general aviation in 1980 was 5 times safer than the commercial sector.

However, the number of fatalities or injuries per hours flown is not a very reliable measure of safety if we just extract the figures for one year, so it would therefore be necessary to use the figures from a greater number of years for a justifiable comparison to be made. Anyone who cares to do the arithmetic necessary, and using the figures given in the CAA annual surveys *Accidents to Aircraft on the British Register* from 1960 to 1982, will find that the total hours flown by public transport/commercial services came to approximately 17 million, during which 1,681 fatalities occurred, i.e. 1 fatality for every 10,113 hours flown. During the same period the CAA annual surveys estimated that general aviation flew approximately 10.5 million hours, during which time 534 fatalities occurred, i.e. 1 fatality for every 19,663 hours flown (Fig. 1.4).

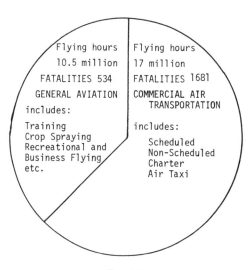

Fig. 1.4

It could thus be argued that, as far as the UK is concerned, when it comes to the measure of safety in human terms and in relation to their respective flying activities, general aviation is almost twice as safe as the airline passenger services (Fig. 1.5).

However, to measure the aviation risk factor in terms of flying hours versus number of fatalities is to ignore the known fact, mentioned earlier, that over 65% of all accidents to aircraft occur during the take-off and initial climb and the approach and landing phases of flight. Hence the number of flights rather than the number of hours flown would be a more sensible basis upon which to carry out measurements of safety rates. This aspect was pointed out in an article concerning world airline safety in the January 1985 issue of *Flight International*.

If we were indeed to use this method then the general aviation safety

HOW SAFE IS GENERAL AVIATION?

UK CIVIL AVIATION

1960 to 1982

COMPARISON OF HOURS FLOWN PER PERSON KILLED

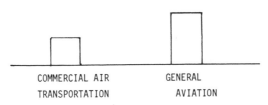

Fig. 1.5

factor would be further improved. This can be determined by referring to the ICAO *Statistical Year Book* in which the average stage flight by aircraft engaged in scheduled air transportation services is reported as 1.43 hours, and an average stage flight by aircraft engaged on non-scheduled commercial air transportation is shown as 2.05 hours. Therefore if we use the lower of these two figures to allow for the effect of air taxi operations, then taking the 17 million hours done by all commercial air transportation between 1960 and 1982 we see that there were approximately 12 million stage flights. If the total of 1,681 people killed during these operations is now divided into this figure it works out at 7,138 flights per fatality.

UK CIVIL AVIATION

STAGE FLIGHTS PER PERSON KILLED 1960 to 1982

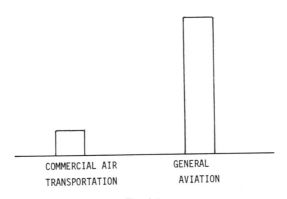

Fig. 1.6

Turning to the record of general aviation during the same period, it becomes difficult to obtain an accurate factor for converting the flying hours done into stage flights, but by the very nature of general aviation activities and the smaller size of aircraft it would be reasonable to assume that an average stage flight would be in the region of 30 minutes (bear in mind that flying training forms a large part of general aviation activities). If this factor is used in relation to the 10.5 million hours carried out between 1960 and 1982 then the number of stage flights would be 21 million and this works out at 39,326 flights per fatality (Fig. 1.6). On this basis general aviation, during the 22-year period, would appear to have been 5.6 times safer than the commercial air transportation sector.

Similar figures for general aviation in other countries are given in Tables 1.1 and 1.2. Although these are based on shorter periods it would appear that these countries have a better record than the UK in relation to the safety of persons per hours and stage flights flown.

Table 1.1 General aviation, Australia

Year	Accidents	Fatalities	Hours flown	Hours per fatality
1981	173	42	1,175,600	27,990
1980	192	29	1,194,800	41,200
1979	183	25	1,124,800	44,992
1978	202	40	1,067,600	26,690
1977	177	30	1,075,900	35,863

5-year average 1977 to 1981

Hours flown per fatality	33,968
Approximate number of stage flights per fatality	67,936

Table 1.2 General aviation, USA

Year	Fatalities	Hours flown	Hours per fatality
1981	1,208	36,803,000	30,466
1980	1,180	36,402,000	30,849
1979	1,169	38,641,000	33,054
1978	1,487	34,887,000	23,461

4-year average 1978 to 1981

Hours flown per fatality	29,090
Approximate number of stage flights per fatality	58,180

Finally, in making comparisons between the safety record of general aviation and the air transport sector we also have to bear in mind the facts given in Table 1.3. The comments in this table refer to aviation activities in the UK, but are probably broadly true for many other countries.

Table 1.3

- 90% of pilots used in commercial air transportation have an instrument rating.
- 2% of private pilots have an instrument rating.
- 75% of the hours flown in commercial air transportation are carried out with two pilots on board.
- 95% of the hours flown by private pilots are conducted with only one pilot on board.
- 90% of aircraft used in the commercial air transportation sector have two or more engines.
- 90% of aircraft flown by private pilots have only one engine – a significant factor when one engine fails.
- Pilots involved in commercial air transportation fly on average 250 to 300 hours per year.
- Private pilots fly on average, say, 10 hours per year.
- Pilots involved in commercial air transportation have on average 5 times more training in terms of hours and experience than private pilots.
- 85% of the flights undertaken in commercial air transportation take place in the confines of protected airspace.
- 90% of the flights undertaken by private pilots take place in uncontrolled airspace.
- Commercial air transportation flights are conducted mainly from large and well-equipped aerodromes.
- Private pilots operate from small aerodromes and landing strips, the latter leaving little margin for pilot error.
- 60% of all accidents occur during the take-off and initial climb, and the approach and landing phase of the flight:
    The average length of an air transportation flight is 2 hours.
    The average length of flights conducted in general aviation is 30 minutes.
  Thus general aviation flights are exposed to the highest risk area 4 times as often than those in commercial air transportation.
- Additionally the annual total flying hours carried out in general aviation include some 200,000 hours with student pilots at the controls.

The safety record reworked in accordance with the foregoing comments is however not intended to imply that flying in an airliner is more dangerous than flying in a light aircraft or vice versa, nor are the figures intended to imply that flying is more risky than crossing the road or driving in a car. Rather, the figures show how care must be used when interpreting statistics of any sort. For example, the statistics for UK commercial air transportation during the years 1981, 1982 and 1983 (the latest available at the time of printing) show that during this period the number of stage flights per fatality rose to 26,221, an impressively better record than for the previous 22 years and one which reflects the overall improvement made with regard to aviation safety in the air transport role. Against this is the fact that the year

1985 was a particularly bad one for the commercial air transport sector and was described by the media as 'the worst year in aviation history'. Emotive news reporting of this nature may be good for selling newspapers but allows nothing for the fact that year after year aviation safety has its variations whilst at the same time the number of passengers carried increases steadily, a factor which has to be taken into account when any reflections on aviation safety are made.

In summary, it should be appreciated that in aviation there is the important fact that although accidents do not automatically cause injuries or death, injuries or fatalities normally only occur as a result of an accident. Therefore whatever the ratio between accidents, flying hours, injuries or fatalities it is clear that one accident is one too many and we would all be safer if the number of accidents were reduced. This philosophy is equally true whether we are flying in the airline passenger section where a single accident could cause the deaths of many, or in the general aviation sector where the occupants of an aircraft are normally few in number. No pilot will obtain solace from knowing that one sector of aviation is less safe than his own when he, together with his passengers, is dead or lying injured as a result of an accident, whether it be in aeroplane or any other form of transport.

Now, having shown the need to interpret statistical evidence with care, it can also be said that the correct interpretation of accident statistics can be of considerable value in attempting to reduce the risks during flying because the past figures show those areas which have led to the least safety in numerical terms. Another important message gained from examining past accidents is that very few accidents occur as a result of the pilot not knowing how to handle the flying controls. What does stand out quite clearly is that the cause of most accidents has been a 'lack of good judgement' rather than physical handling skills. In this respect it should be understood that although there is a degree of constancy in the yearly frequency of certain causal factors in general aviation accidents, it is equally true to say that the extent of the pilot's role in these accidents increases from year to year, and for this reason it would be useful to consider the changes to the pilot's total environment and his operating problems if we are to understand the reason for the predominance of 'pilot error' and if we are to improve the safety factor in our general aviation activities.

The development of general aviation

In 1962 the UK general aviation fleet was composed of approximately 1,300 aircraft, mostly of simple construction and fitted with very basic equipment. In 1984 the fixed wing and helicopter segment of the UK general aviation fleet below 2,300 kg had grown to over 6,000 aircraft, many of these being fitted with very sophisticated equipment and systems. The range of speeds between the slowest and the fastest aircraft has increased to a considerable extent, and where once the speeds of all general aviation aircraft were

Fig. 1.7

similar, today we have small trainers using the same airspace as jets, and this is an extremely significant factor, particularly in the mix of aircraft within aerodrome traffic areas.

Fig. 1.8

In 1962 only a small number of the GA fleet operated in accordance with IFR but today, with the advent of improved flight instruments and radio navigation equipment and the reliability of avionics, even the smallest aircraft of the fleet commonly operate in weather conditions less than VMC. The air traffic system has also developed both in sophistication and technology and the attendant increase in the body of regulations and procedures, etc., which the pilot has to accommodate whilst carrying out the basic function of flying the aircraft gives additional problems to the general aviation pilot.

It is also pertinent to observe that the airspace system with its many complications (Fig. 1.9) has been developed primarily for the benefit of the

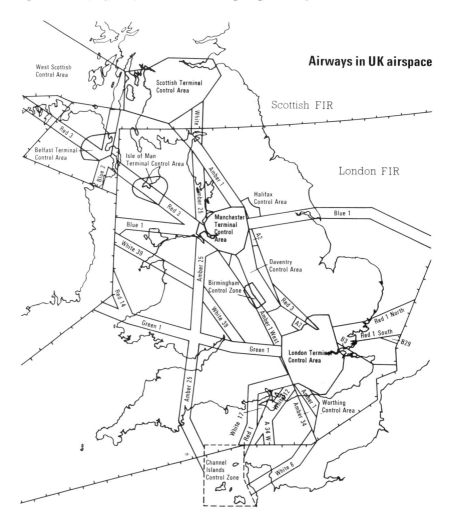

Fig. 1.9 The need to avoid controlled airspace adds considerably to the private pilot's workload during flight.

safety and smooth operation of large passenger-carrying aircraft which may have two or more skilled and full-time professional pilots on board. Thus single pilot operation has often only been a secondary consideration during the development of the air traffic control environment, and by the very

HOW SAFE IS GENERAL AVIATION?

nature of things this is also one in which air traffic controllers have problems, in that they are mostly unaware of the flight experience of the pilots they are controlling or giving instructions to.

All these factors have combined over the years to produce a substantial increase in the pilot's workload. A pilot's safety and that of his passengers is directly related to his competence, measured in terms of physical skills and judgement. Whenever task loads are high, both of these factors can be adversely affected. It is hoped that the information contained within the following pages will enable a pilot to develop better judgement through a greater understanding of the problems involved in operating an aircraft with safety, by increasing his knowledge and awareness. This will improve his **cognitive ability**, which simply means '**putting it all together and making the right decision at the right time**'. It is this ability which appears to have been missing in a large number of the 3,740 accidents to aircraft on the British register in the past 20 years.

> **In no sector of aviation does the responsibility for the safety of his flight operations devolve more personally than on the general aviation pilot who so often operates independently of any form of supervisory control.**

Fig. 1.10

2
Motor skills and human factors

It could be said that in the very early days of aviation, every time an aeroplane staggered into the air a state of emergency existed. The incidents and accidents which occurred during this period were in the main caused by structural or mechanical failure. These events usually happened within a short time of the aeroplane becoming airborne and as such the pilot rarely had enough time to make a mistake.

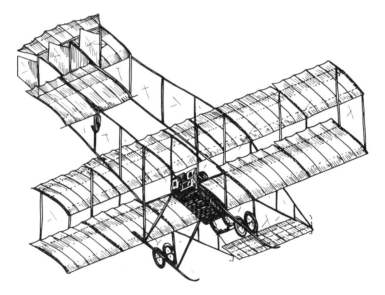

Fig. 2.1

Later on, as aviation developed, the majority of the accidents became attributable to pilot error, though to be fair to the pilot, a number of these errors were compounded or even triggered off by mechanical or systems failure. It would nevertheless appear that in most cases pilot error is primarily caused by such factors as lack of preparation, inadequate training, or an inability to make the correct decision and take positive action at the appropriate time.

Today, the aircraft we fly have very high standards of reliability – the results of the progress made in aviation technology and engineering. This

MOTOR SKILLS AND HUMAN FACTORS

aspect has been assisted in no small way by the dedicated efforts of aircraft accident inspectors whose work, particularly in the more advanced aviation countries, has been an added incentive to aircraft manufacturers and aviation authorities in achieving improvements to the overall safety of aviation activities.

Today, pilots receive training to far higher standards than in days gone by. This training enables them to develop motor skills for controlling the aircraft and its systems. Before a pilot can qualify for a licence today he has to undergo tests in both knowledge and skill, yet current accident records clearly reveal that in the main it is the pilot who, through basic errors and misjudgement, causes the largest percentage of incidents and accidents.

Fig. 2.2

Why? Perhaps it would help to explain this if we listed the two main areas involved in the determination of pilot competence. These are:

(1) The acquisition of motor skills.
(2) The development of decision-making ability.

When considering these aspects in relation to reducing the risk factor it would be useful to establish whether one or the other of these two qualities has been shown to be the weakest link in the chain of events.

On the surface it would seem that lack of flying skill in the physical sense should not feature too widely because the qualified pilot has had to demonstrate a satisfactory standard of competence in controlling the aircraft and also an acceptable knowledge of the technical subjects in the pilot training syllabus. This premise has nevertheless to be accepted with some reservation because a pilot may not have flown for some time prior to the accident, or may have lacked experience on the aircraft type, or in the environmental conditions in which the aircraft was being flown. Added to this is the fact that any pilot who is out of practice or operating in an unfamiliar environment, or lacks recall of essential information, will automatically have a reduced judgement and decision-making capability. From this it can be seen that on those occasions when physical flying skills are lacking it is also likely that the ability to make sound judgements and carry them out will also be lacking.

However, in acknowledging this fact it must be appreciated that many incidents and accidents occur to pilots who are in continuous flying practice and who have to recall their essential aviation knowledge on numerous occasions. The point which could be made in this case is that these pilots will be at risk in the aviation sense more frequently and therefore will be likely to feature more often in the accident statistics. Appreciating this fact, however, will do nothing to reduce the accident rate, nor will it determine the best way to reduce the risk factor.

On a related point, it has already been said that mechanical or structural failure is the least direct cause of today's accidents in aviation but in considering this fact it has also to be borne in mind that a number of minor unserviceabilities to the aircraft equipment and systems still occur, and whereas these in themselves may not be serious they can increase the pilot's workload and create distractions to the extent that the pilot's decision-making qualities are weakened. They will usually also increase the frequency of the decisions to be made.

Armed with these thoughts it would be useful to research the available reports on aircraft incidents and accidents to see if there is a principal underlying cause of human errors occurring. If we could pinpoint a common cause then we would also have an opportunity to reduce the risk factor every time we fly. However, it is not the intention to produce reams of accident statistics in this book, but rather to highlight certain facts and bring them to your attention.

In reviewing the accident summaries over the past 10 years one significant fact emerges – and that is that the lack of motor skills, i.e. the ability to control the aircraft and its systems, played a very small part in the number of accidents which occurred. On those occasions when lack of skill was evident it was mainly related to such actions as inadvertently selecting the wrong fuel tank, forgetting to lower the landing gear, and similar. In other words it was not so much a case of the pilot not knowing what to do with the controls, but rather a case of **not doing it, or doing it incorrectly**.

So, few accidents occurred because the pilot did not know how to control the aircraft and its systems. On the other hand, what does appear from the review is the fact that due to distractions, or for other reasons, the pilot failed to use correct judgement. To give a typical example of this as far as landing accidents were concerned: there were many occasions when the aircraft was perfectly serviceable, the weather was fine, the pilot was in current flying practice on the aircraft type and the landing was being made on an aerodrome which had been used by the pilot on frequent occasions, yet the aircraft finished up by running out of runway and crashing through the far hedge*... These accidents occurred to qualified pilots with total flying hours ranging from hundreds to thousands.

Fig. 2.3

* This type of accident is particularly prevalent when pilots are using shorter grass runways at aerodromes and landing strips.

The pilots concerned had all demonstrated their ability to land safely on hundreds of occasions and there were no mitigating circumstances, therefore one can only conclude from this that they were victims of the compulsion brought about by pressures of pride or time, or complacency or carelessness, as all these accidents could have been avoided if the pilots had made the sensible decision to *go round again*.

Fig. 2.4 (Photograph courtesy of the Australian *Aviation Safety Digest*)

The records also show that many aircraft failed to become airborne in the available length of the take-off run and finished up in the far hedge, or stalled just outside the aerodrome boundary.

It would therefore appear that compulsion, carelessness or complacency must figure very high up the scale when accidents of this sort take place. In addition there may also be an element of lack of knowledge relating to the take-off and landing performance of the aircraft, which the pilot has either not been taught or has failed to understand. Thus any pilot who can programme his flying activities with these possible shortcomings in mind and who is sensible enough not to fall victim to these human failings will already have done a great deal to reduce the risk factor.

Nobody is born with the ability to fly, it is a task which can only be learned through education and practice. It has already been stated that this task can be divided into two distinct parts, the development of motor skills and the ability to make good decisions. In the early stages of learning it is logical to concentrate upon the acquisition of motor skills and simple judgement, the latter being generally confined to such items as correctly positioning the aircraft in the traffic pattern, using the flying controls and throttle to achieve the correct approach path and speed, using the controls and judging the height of the aircraft during the landing flare, etc., and the development of more complex judgements based on multiple factors is left until a later stage in training is reached.

Fig. 2.5 (Photograph courtesy of the Australian *Aviation Safety Digest*)

In the initial stages of training, the ability to handle the aircraft and make it do what the pilot wants appears to be the most difficult of the two parts to both the student and the instructor, whereas in reality it is the easiest part of learning how to operate an aircraft with safety. The basic reason for this is that motor skills are developed by handling items which can be seen, felt and moved. On the other hand, judgements and decisions are more abstract and are based upon processes using intelligence and awareness.

Fig. 2.6

In summary, it is clear that motor skills are not enough, nor indeed is knowledge. Once a pilot has developed the physical skill to operate aircraft and once he has acquired the technical knowledge needed he will still have to develop the cognition so necessary for good judgement. The word cognition is used without apology because it encompasses the process of making judgements and decisions based upon a sense of awareness gained by knowledge. To put this another way, it is no use just learning physical skills or memorising items of knowledge if you wish to reduce the risks in flying. The first of these only permits you to enter a further risk environment – the air ... The second – knowledge – will only be of value if it is used properly, i.e. to come to a correct or wise decision. It is a fact that many pilots of undoubted skill have lost their lives by getting into situations where skill alone could not extricate them, and many pilots who have gained first-class marks in written examinations have gone the same way. So although both skill and knowledge are necessary they will not by themselves create a safe pilot. The ingredient which has to be added is good judgement leading to correct decision-making.

High-risk phases of flight

If we accept that incorrect decisions on the part of the pilot lead to the greatest number of accidents, the next step should be to find out in which phases of flight the highest risk factor is produced, i.e. where most accidents happen. The answer to this will indicate where good decision-making by the pilot is at its weakest. It will also be useful to plot the average time of exposure to each phase.

Table 2.1

UK fixed wing aircraft, non public transport, 5-year period		
Flight phase	Average time spent per phase (minutes)	Number of accidents
Taxying and run-up	7	86
Take-off and initial climb	2	159
En-route climb	5	18
Cruise	19	70
En-route descent	5	16
Initial approach	3	15 ⎫
Final approach and flare to touchdown including the landing run	2	411 ⎬ 487
Touch and go	1	31
Go-around	1	30 ⎭
Assumed average length of flight: 45 minutes		

The total number of accidents in the above phases of flight for the 5-year period was 836. Accidents which involved aerobatics, display flying, and crop spraying were not included.

However, whereas the allocation of accidents to flight phase is simply obtained by looking at the records, the average length of a flight in terms of time can only be estimated. Nevertheless the length of time spent by an aircraft during, for example, take-off and initial climb, and final approach to touchdown, can be realistically assessed as shown in Table 2.1. It is more difficult to assess the time spent in cruising flight.

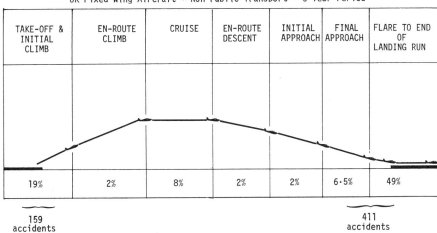

Fig. 2.7

Examining the figures in Table 2.1 we see that in the two 2-minute periods of 'take-off and initial climb' and 'approach and landing' the greatest exposure to risk occurred: in this 4-minute band of flight 570 of the 836 accidents took place. Fig 2.7 shows the degree of risk by phases of flight in terms of percentage of the total accidents shown.

Although comparisons using a single year of operation have little validity for statistical purposes it is nevertheless interesting to note the similarity which exists on a more international level in relation to the percentage of accidents which occur in the various flight phases. Table 2.2 shows the percentage of accidents per phase of flight for several countries during 1981.

Table 2.2 Accidents by phase of flight, 1981 (per cent)

Country	Take-off and initial climb	En-route climb	Cruise	En-route descent	Initial approach	Final approach	Flare to end of landing run
UK	16	1.5	0.5	1.9	2.4	6.5	43
USA	20	2.3	13.1	2.4	2.5	7.4	32.8
Australia	18.5	3.4	9	1.3	–	6.9	43.8

Another way of comparing the risk by phases of flight is to examine the accident figures in relation to the numbers of accidents which occur on, or very close to, an aerodrome. Tables 2.3 and 2.4 show extracts from the statistics produced by the Bureau of Air Safety Investigation in Australia and the National Transportation Safety Board in the USA. In Table 2.3 the total of all accidents which occurred within $1\frac{1}{2}$ kilometres of an aerodrome adds up to 65% of the total accidents for the year. In Table 2.4 we see that the number of accidents which occurred within 1 mile of an aerodrome add up to 61% of the total.

The most common causal factors

Having established the degree of risk in relation to the flight phase the next step is to determine the types of accident which most commonly occur in each phase. Once this is done a review of the circumstances leading up to the type of accident can be made in an attempt to analyse why these accidents occur so frequently.

It has been seen that the flight phases in which the highest risk of an accident occurs are the take-off, en-route and the final approach and landing. Table 2.5 lists the most common causes of general aviation accidents which occur in these phases.

Although weather-related accidents are not as common as might be expected they have been included in the above list because the results are so often fatal. On another aspect, about one-third of the total flying hours carried out in the UK general aviation section each year is devoted to training, yet the student pilot is rarely involved in an accident. This point is made because in considering the landing phase accidents one could be excused for thinking that the majority of hard landings and similar accidents would occur with a student at the controls, but of the 43 landing accidents in the UK in 1982, only 7 involved students and of the 83 landing accidents in 1981 only 9 involved students.

One problem which arises when attempting to determine the primary cause of an accident is the difficulty of obtaining the complete train of events which led up to it. Due to the law and litigation there is a natural human reaction which may prevent an investigator obtaining completely true facts, added to which there are the after-effects on the human memory when the accident was of a traumatic nature.

Another problem in the UK is that the general aviation sector is relatively small; for example, the flying hours it carried out in 1983 were approximately 700,000, whereas in the USA in the same year, the general aviation activity accounted for more than 35 million hours. The fewer hours done in the UK mean fewer accidents from which to examine causal factors, and therefore it is often necessary to refer to statistics produced in countries where aviation is practised more actively. This can be done without spoiling the validity of the answers we need because in these countries the aeroplanes are the same, the man-machine environment is the same and the life-style of the pilots is sufficiently similar to enable lessons to be learned. Added to

Table 2.3 Selected accident data, all operations, 1981
Airport proximity by highest degree of injury, fatalities, fire after impact and damage (excluding gliding)

Aerodrome proximity	Total	Highest degree of injury			Aircraft damage			Number of fatalities	Fire after impact	
		Fatal	Serious	Minor/nil	Destroyed	Substantial	Minor/nil		Fatal	Non-fatal
On aerodrome	117		3	114	2	114	1			
On heliport	2			2	1	1				
Within 400 m	21		2	19	4	17				
Within 800 m	6	1		5	4	2		3		1
Within 1 km	5		1	4	1	4				2
Within 1½ km	4	1	1	3	2	2		1	1	1
Within 3 km	4	1		2	2	2		1	1	
Within 5 km	3	1		2	2	1		7	1	
Within 6 km	2			2	1	1				
Within 8 km	2	1		1	1	1				
Beyond 8 km	25		3	22	6	19		4		
Not relevant/unknown/not reported	44	13	3	28	16	27	1	32	4	
Totals	235	18	13	204	42	191	2	48	7	4

Table 2.4 Aircraft by proximity to airport and flight plan, all fixed wing aircraft, 1981

Proximity to airport	Flight plan									Aircraft	
	No flight plan	VFR	IFR	Controlled VFR	IFR (VFR condition on top)	DVFR	VFR flight following	Special VFR	Unk/NR	Other	No.	Per cent
On airport	1154	109	65	0	0	0	0	2	15	3	1348	42.4
On seaplane base	12	1	0	0	0	0	0	0	0	0	13	0.4
In traffic pattern	179	19	5	0	0	1	1	0	0	1	206	6.5
Miles from airport:												
Within ¼	149	3	7	0	0	0	0	0	2	0	161	5.0
¼ to ½	93	11	4	0	0	0	0	0	0	0	108	3.4
½+ to ¾	19	0	5	0	0	0	0	0	0	0	24	0.8
¾+ to 1	75	5	9	0	0	0	0	0	0	0	89	2.8
1+ to 2	94	12	17	0	1	0	1	0	1	1	127	4.0
2+ to 3	50	6	9	0	0	0	0	0	0	0	65	2.0
3+ to 4	29	2	8	0	0	0	0	0	0	0	39	1.2
4+ to 5	20	3	4	0	0	0	2	0	2	4	27	0.8
Beyond 5	636	90	56	1	0	0	1	0	2	4	791	24.8
Unknown/not reported	161	13	11	0	0	0	0	0	7	0	193	6.0
No. of aircraft	2671	274	200	1	1	1	5	2	27	9	3191	
Per cent	83.7	8.6	6.3	0.0	0.0	0.0	0.2	0.1	0.8	0.3		

Table 2.5

Flight phase	Cause of accident
Take-off	Inadequate distance Wind gusts Loss of control on ground Loss of control in flight Engine malfunction
En-route	Engine malfunction Weather-related
Approach and landing	Loss of control in flight Engine malfunction Inadequate distance Gear up Hard landing Gear collapsed Nose over/down Over-run Collision with objects Loss of control on ground

which is the fact that the substance of the definition of a reportable accident as used by the CAA and the FAA is the same. In this respect an examination of the FAA accident statistics over a period of years shows that the percentage of accidents in the USA per phase of flight is closely similar to those of the UK.

The National Safety Transportation Board of the United States revealed in a recent review that in accidents where fatalities occurred, 7 out of 10 of

Fig. 2.8

the leading causal factors involved some type of human error, while the remaining 3 involved environmental conditions. In taking the top 10 leading causal factors by frequency of occurrence for one recent year it was revealed that human error was a causal factor in all cases.

Conditions which aggravate pilot error

It would nevertheless be an oversimplification of the facts of life if one were to simply blame the pilot for causing the majority of accidents. Whilst there are a number of cases where the blame could be attributed entirely to the pilot, there are many other cases where this does not hold true.

For example, the pilot has to operate in an environment dictated not only by the weather but also by man's technology and the imposition of regulations. If technology makes unnecessary difficulties for a pilot, if regulations do not take into account the limitations of human memory, particularly when carrying out the many airborne tasks demanded of a pilot, then these too must accept part of the blame when a pilot is placed in overdemanding situations and as a result has an accident.

A few years ago the AOPA Flight Safety Foundation in America commissioned a study to review the current status of general aviation safety with a view to providing a logical basis for a renewed, possibly expanded, and highly focussed programme of efforts designed to significantly reduce general aviation accidents and the resultant injuries and fatalities. The following extracts from the final report of this study are pertinent to the aspect of pilot error:

(1) It is clear from the accident reports that the root of the problem is the dominance of pilot error as a cause or factor in accidents. However, it is the view here that just laying the blame on the pilot is simplistic. While there are frequently cases where this is appropriate, there appear to be many others where it is not. In the latter case it appears that the pilot is the victim of one or more accidents of history, bureaucracy or evolution, and is led to pilot error by circumstances that overload his ability to perform multiple tasks simultaneously under a body of instructions, regulations and prescribed procedures that defy any but a 'Philadelphia lawyer' to understand and remember. The system has not been designed to meet his needs and safety, and neither has the aircraft...

(2) A study of aircraft 'design-induced pilot error' reveals a number of ways in which the aircraft design or operations manual either induced pilot error or made it difficult for the pilot not to commit an error. A trip to the flight line today will provide confirmation that the condition still prevails. There are such things as lack of standardization of controls, ill-conceived fuel management systems, poorly placed or nearly inaccessible controls, fuel filler openings that require obtaining a stepladder to visually check the contents of the tanks, adjustable crew seats that fail to engage and lock in place, methods of systems operation which when

used must distract the pilot from other vital functions, and many more.

It is conceded that under normal circumstances, or with highly seasoned pilots, none of these might be injurious; but study of accident records shows that quite frequently the circumstances are not normal and the pilot is not highly seasoned, and to reduce the accident rate among the less experienced pilots, these undesirable features should be removed. If they are removed it will also improve the accident rate among experienced pilots as well...

There is no visible case where making the control of the aircraft easier and less demanding does not seem to be in the right direction. It is believed that the increase in confidence and reduction of anxiety would allow better concentration on command and the exercise of judgement, and would even influence such accident rates as those associated with improper level off in landing...

There is no doubt that the reasons for pilot error are often due to circumstances which overload his ability to perform multiple tasks simultaneously, particularly as it is a recognised fact that humans have only one decision channel and thus can make only one decision at one time. Therefore tasks which combine a need to understand and recall a wide variety of regulations and operational procedures concerning the management of the aircraft can from time to time create elements of confusion in a pilot's mind, leading to delays in decision-making, or create a situation in which decisions are made in the wrong order, or simply produce the wrong decision.

The comments made in the following sections of this book can do little to simplify technology or reduce the vast number of regulations; nevertheless they attempt to show that with a little thought a pilot can reduce his risk factor through developing a greater awareness of the problems, thus leading to better flight planning and operating habits.

It was stated earlier that judgement is closely linked with an attitude of mind and in this respect it is important to understand that human error is not a stigma; all humans make mistakes, from the time they get up in the morning to the time they go to bed. However, whereas many of these mistakes are relatively unimportant, such as dropping a pen or a spoon, there are others which, if made in certain circumstances such as driving a car or flying an aeroplane, could lead to disastrous consequences. Therefore in the same way that a pilot learns to control his aircraft he must also learn to control his actions in other directions. As a first step he must understand that whereas mistakes will happen, the number of mistakes which could occur can be reduced by intelligent training. Training coupled with self-discipline is therefore an important key to the competence required to achieve flight safety. But we need to know what it is we are training *against* as well as *for*. Perhaps the answer to this can best be summed up by saying that we are training ourselves to combat our own human weaknesses.

To expand on this theme a little further it would be of value to consider how the development of habits fits into the picture. For example, from a very early age we learn to throw stones, and later cricket balls and darts. In each case we learn to aim the object we throw and this indeed becomes a 'habit'. Relating this habit to a flying situation such as a pilot finding himself too low on an approach to land, it becomes clear that no matter how well he has been trained there will be an intuitive, almost subconscious tendency to ease back on the control column and raise the nose of the aircraft. This has been proved by observation to apply to most pilots, but the amount of control movement before application of power will be significantly greater in the case of the inexperienced pilot than one who has thousands of hours in his log book. So whatever a pilot's experience it can be seen that a habit learned in childhood will require the strongest possible mental and physical effort to resist.

Fig. 2.9 (Photograph courtesy of UK Accident Investigation Branch)

> An aeroplane is perfectly safe ... until the pilot climbs on board ... after which anything can happen and often does!

MOTOR SKILLS AND HUMAN FACTORS

So what do we learn from this?

Provided we appreciate that we are all creatures of habit, we are more easily able to guard against those habits. They become more recognisable, and therefore we can where necessary take steps to suppress them, though, alas, it will be difficult, and sometimes in moments of stress almost impossible to overcome them completely.

On the following pages a number of the most common causes of incidents and accidents are discussed under the headings which in the past have proved to be the phases of flight in which serious accidents occur. Suggestions are also made in relation to pilot procedures and habits which if developed and used could reduce the risk factor in your aviation activities.

3
Preflight planning and preparation

Depending upon the purpose of any flight the pilot's preflight planning will vary considerably; for example, the amount of preparation necessary for a short flight in the aerodrome circuit will be significantly less than that undertaken when a cross-country flight is planned.

Nevertheless the considerations which are appropriate to each type of flight have many common features and a review of the causal factors of accidents on a world-wide basis reveals that the lack of, or inadequate, preflight preparation sits high on the causal factor ladder (Tables 3.1 and 3.2). In many countries this cause has consistently topped the list for many years.

Table 3.1 Most prevalent detailed accident causes – all operations, 1981 (Annual review of aircraft accident data, US general aviation)

Detailed cause	Number of accidents	Per cent of accidents
Pilot – inadequate preflight preparation and/or planning	360	10.3
Pilot – failed to obtain/maintain flying speed	339	9.7
Powerplant – failure for undetected reasons	257	7.3
Pilot – mismanagement of fuel –	246	7.0
Fuel exhaustion	197	5.6
Material failure	184	5.3
Pilot – selected unsuitable terrain	170	4.9
Pilot – improper level off	169	4.8
Pilot – misjudged distance and speed	166	4.7
Pilot – continued VFR flight into adverse weather conditions	157	4.5

What is not so easily determined is the method used in different countries to assess this cause in relation to other listed reasons. For example, although an accident may be given the causal factor of inadequate take-off or landing distance, fuel exhaustion, or mismanagement of the fuel system, the initial cause may really have been improper preflight planning. Thus this cause may be far more common than revealed in the relevant tables of publications

PREFLIGHT PLANNING AND PREPARATION

Table 3.2 Survey of accidents to Australian civil aircraft, general aviation operations, 1981

Assigned factors	Student	Private Restricted	Private	Commercial	Senior Comm/ATP	Unknown/other	Total
Pilot factors by type of licence held							
Inadequate preflight preparation/planning			13	25			38
Improper in-flight decisions or planning (includes taxi phase)	1		20	9	1		31
Selected unsuitable area (for take-off, landing, taxi)			17	11			28
Did not see/avoid objects/obstructions (excludes when reasonable precautions taken)			7	20			27
Attempted operation beyond experience level	5	1	12	9			27
Did not obtain/maintain flying speed	3	1	9	9			22
Improper compensation for wind conditions (e.g. drift corrections)	1		14	4	1		20
Diverted attention from the operation of the aircraft	1		4	14			19
Improper landing flare	3	1	15	1			20
Did not initiate go-around	3		13	3	1		20
Improper operation of brakes/flight controls (in ground operation)	1		6	6			13
Lack of familiarity with aircraft (for type of operation attempted)	1		7	4	1		13
Improper recovery from bounced landing	3		10		1		14
Improper operation of primary flight controls (includes trim controls, excludes flaps, etc)	6	1	4	3			14
Misjudged distance and speed (overshoot)			6	4			10
Delayed in initiating go-around		1	6	3			10
Approached high and fast			4	4	1		9
Maintained excessive airspeed		1	7				8
Inadequate supervision of flight (in multi-crew operations)			1	4	3		8
Did not use or incorrectly used miscellaneous equipment			2	5	1		8

relating to aircraft accidents. Because of this, the comments relating to this type of causal factor are integrated, where applicable, into the text of the subject matter throughout this book; and this section is confined to a broad appraisal of what preflight preparation is all about.

Flight planning considerations

Whilst any safety-conscious pilot will only plan to fly within the limitations of his own skill and his aircraft's equipment, it must be appreciated that weather prediction still remains an inexact science and many pilots have found to their embarrassment and sometimes cost, that the forecast weather during a flight has deteriorated to the point where competence in instrument flying has become an essential requirement if the flight is to be continued safely.

In addition to this factor there have been significant changes to the once simple airspace environment within which the early aviator flew. The development of airways, control zones, special rules areas and zones, danger and similar hazardous areas, has brought about restrictions which affect most cross-country and local flights.

Today, the ability of a pilot to navigate accurately and to implement the correct procedures in relation to this type of airspace while operating within his legal privileges, demands an ever-growing need for more competence. The problem increases whenever visibility conditions are less than those acceptable for safe visual flight.

Having said this, it must also be borne in mind that the development of instrument flying skill and the ability to use radio navigation aids does not necessarily make a person an all-weather pilot. The light aeroplane of today has limitations with regard to severe weather, e.g. airframe icing, turbulence, etc., and in relation to airframe icing it must be realised that on many days

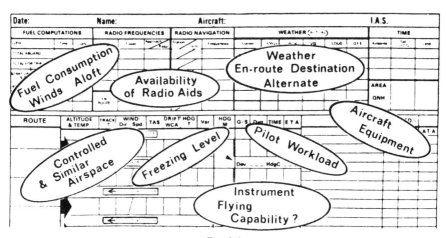

Fig. 3.1

during the winter months the freezing level will coincide with the lower levels at which light aircraft normally operate. Very few light aircraft, whether single- or multi-engined, are fitted with equipment which permits them to be flown in airframe icing conditions, and therefore when airframe icing conditions exist, it will usually be both hazardous and illegal to conduct flights in cloud.

Because the weather conditions which prevail or which are anticipated from the time of take-off and during the en-route and arrival phases are vitally important to the pilot's determination of whether the flight can be safely conducted, a thorough understanding of the sources of meteorological information available to the pilot, and an ability to interpret and assess this information, are a positive requirement.

Most pilots are reluctant to admit to being disorientated or lost. Being lost can be an embarrassing, and sometimes frightening, experience. The most common causes of becoming lost, an occurrence which so often starts the chain of events which lead to a hazardous situation or an accident, are listed below:

(1) Becoming lost after encountering poor weather/visibility.
(2) Poor use of radio/navigation aids.
(3) Poor flight planning.
(4) Failure to obtain clearance and poor knowledge of procedures.

A detailed analysis of these causes showed the following results:

- Poor visual navigation included:
    poor map reading
    misidentification of towns
    incorrect identification of main roads and railway lines
    wind error not noticed due to bad map reading
    drift not adequately compensated/poor estimation of drift
    failure to re-synchronise heading indicator with the magnetic compass.
- Poor use of radio/navigation aids included:
    flying QDR instead of QDM
    no cross-checking for heading indicator precession
    following incorrect VOR radial
    following VOR only and not allowing for possible navaid errors
    none or poor cross-checking of navigation aids or navigation information
    failure to select another VOR after finding a VOR was not working.
- Other, general causes included:
    not carrying adequate maps
    out of date charts/maps
    distraction during radio trouble
    flying wrong headings from turning point

overshooting reporting point
turning too early
misreading protractor during flight planning
putting wrong radio frequency in navigation log.

One of the actions most neglected by a pilot contemplating a flight is that of properly completing the preflight planning procedures – a fact which is well supported by the previous accident statistics. The following elements are vital during preflight preparation:

The selection and use of charts

A basic element of preflight preparation requires the use of appropriate and current navigational charts on which pilots can mentally review their intended route of flight. The use of outdated charts may result in flights into airport traffic areas, control zones, or restricted areas without proper authorisation. Having available the information contained in current charts will enhance a pilot's ability to complete the flight with greater confidence, ease, and safety. You should examine the projected path across the face of the chart for the location of good check points, restricted areas, regulated and controlled airspace, obstructions, other flight hazards, and suitable aerodromes, etc.

The 1:500,000 aeronautical topographical charts meet the requirements for VFR flights, but suitable instrument procedures and radio navigation charts should also be used when IFR operations are to be conducted.

Another popular chart for longer journeys is the world aeronautical chart (WAC). The scale of the WAC is 1:1,000,000, or 16 miles to the inch. Many states print aeronautical charts which are excellent for VFR navigation within their state boundaries. You should realise, however, that all these charts are designed primarily for VFR navigation and contain only limited information concerning radio aids and frequencies. Thus you should refer to the relevant airman's information publications for more precise coverage of this information.

Planning the route

It is a fact that the shortest distance between two points is a straight line. Consequently, a majority of pilots desire direct routes for most flights. Quite often there are factors to be considered that may make a direct flight undesirable. Restricted and prohibited areas present obstacles to direct flights. In single-engine aircraft, pilots should give consideration to circumnavigating large, desolate areas. With regard to terrain elevation and multi-engine aircraft you should bear in mind the hazards of planning to fly over mountainous areas where the terrain elevation is close to, or higher than, the single-engine service ceiling.

PREFLIGHT PLANNING AND PREPARATION

Fig. 3.2

Aeronautical information publications

The appropriate sections of these publications should be studied during preflight planning. Essential information should be extracted and written on the working flight plan log. Particular attention should be paid to the current NOTAMS.

Weather reports and forecasts

Review weather maps and forecasts, current weather reports, winds aloft forecasts, pilot weather reports, Sigmets, Airmets, and other information. The weather information should be weighed very carefully in considering the go/no-go decision. This decision is the sole responsibility of the pilot and compulsion should never take the place of good judgement.

The navigation log

Precise flight planning of log items, such as precomputed courses, time and distance, navigational aids, and frequencies to be used, will make en-route errors in these items less likely. Special attention should be given to fuel requirements, keeping in mind the need for an ample reserve as well as the location of suitable aerodromes en route where refuelling can be carried out.

Filing a flight plan

The requirement to file a flight plan varies from country to country, and according to the type of flight. However, when the flight is planned to cross desolate areas of mountainous terrain, you owe it to yourself and your passengers to file a flight plan. Taking this action not only enables prompt search and rescue action in the event that the aircraft becomes overdue or missing, but it also permits en-route stations and the destination station to render better service by having prior knowledge of your flight. It costs only a few minutes of time to file a flight plan and may be the best investment you ever make.

Aircraft manual information

Aircraft manuals contain operating limitations, performance, normal and emergency procedures, and a variety of other operational information for the particular aircraft. Traditionally, aircraft manufacturers have done considerable testing to gather and substantiate the information in the aircraft manual. You should become familiar with the manual and be able to refer to it for information relative to a proposed flight.

Weight and balance

The maximum allowable gross weight is established for an aircraft as an operating limitation for both safety and performance considerations. The gross weight is important because it is a basis for determining the take-off distance. If gross weight increases, the take-off speed must be greater to produce the greater lift required for take-off. The take-off distance varies with the square of the gross weight.

Operations within the proper gross weight limits are outlined in each operator's manual. Gross weight and centre of gravity (CG) limits should be considered during preflight preparation. Weight in excess of the maximum certificated gross weight may be a contributing factor to an accident, especially when coupled with other factors which adversely affect the ability of an aircraft to take off and climb safely. These factors may range from field elevation of the airport to the condition of the runway. The responsibility for considering these factors before each flight rests with the *pilot*.

A pilot must not only determine the take-off weight of the aircraft, but must also ensure that the load is arranged to fall within the allowable CG limits for the aircraft. Each aircraft manual provides instructions on the proper method for determining if the aircraft loading meets the balance requirements. The pilot should routinely determine the balance of the aircraft since it is possible to be within the gross weight limits and still exceed the CG limits.

An aircraft which exceeds the forward CG limits places heavy loads on the nose wheel and, in conventional landing gear aeroplanes, may, during braking, cause an uncontrollable condition. Furthermore, performance may be significantly decreased and the stall speed may be much higher.

An aircraft which is loaded in a manner which causes the CG to exceed the aft limit will have decreased longitudinal stability which will give control difficulties, particularly at slow airspeeds. This condition can produce sudden and violent stall characteristics and can seriously affect recovery from the stall. The aircraft will also be much more susceptible to overstress by any control movements or when encountering rough air.

Pilots exceeding CG limits in helicopters may experience insufficient cyclic controls to control the helicopter safely. This can be extremely critical while hovering downwind with the helicopter load exceeding the forward CG limit.

Ice and frost

Ice or frost can affect the take-off performance of an aircraft significantly. Pilots should never attempt take-offs with any accumulation of ice or frost on their aircraft. Most pilots are aware of the hazards of ice on the wings of an aircraft. The effects of hard frost are much more subtle. This is due to an increased roughness of the surface texture of the upper wing and may cause up to a 10% increase in the stall speed. It may also require additional speed to produce the lift necessary to become airborne.

Once airborne, the aircraft could have an insufficient margin of airspeed above the stall such that gusts or turning of the aircraft could result in a stall. Accumulation of ice or frost on helicopter rotor blades results in potential rotor blade stalls at slower forward airspeeds. It could also result in an unbalanced rotor blade condition which could cause an uncontrollable vibration.

Density altitude

This is a very important consideration before take-off. Density altitude is determined by compensating for temperature and pressure variations from the International Standard Atmosphere conditions. A pilot must remember that as density altitude increases, there is a corresponding decrease in the power delivered by the engine and the propellers or rotor blades. For aeroplanes this may cause the required take-off roll to increase by up to 25% for every 1,000 feet of elevation above sea-level. The most critical conditions of take-off performance are the result of a combination of heavy load, unfavourable runway conditions, light winds, high temperatures, high aerodrome elevations, and high humidity.

The proper accounting for the pressure attitude (field elevation is a poor substitute) and temperature is mandatory for accurate prediction of take-off data. The required information will be listed in the aircraft manual and should be consulted before each take-off, especially if operating at a high density altitude or with a heavily loaded aircraft.

Effect of wind

Every aircraft manual gives representative wind data and corresponding ground roll distances. A headwind which is 10% of the take-off airspeed will reduce the no-wind take-off distance by 19%. A tailwind which is 10% of the take-off airspeed, however, will increase the no-wind take-off distance by about 21%.

Although wind direction is a basic consideration to a successful take-off, the number of accidents involving the selection of the wrong runway for the existing wind and taking off into unfavourable wind conditions indicates a need for many pilots to re-evaluate their preflight planning to ensure that the effect of wind is considered fully.

Runway conditions

Aerodromes have various surface compositions, slopes and degrees of roughness. Take-off acceleration is directly affected by the runway surface condition and, as a result, it must be a primary consideration during preflight planning.

Most aircraft manuals list take-off data for level, dry, hard-surfaced runways. The runway to be used, however, is not always hard-surfaced and level. Consequently, pilots must be aware of the effects of the slope or gradient of the runway, its composition, and the condition of its surface. Each of these can contribute to a failure to obtain/maintain a safe flying speed.

The effective runway gradient is the maximum difference in the runway centreline elevation divided by the runway length. Aviation authorities recognise the effect of runway gradient on the take-off roll of an aircraft and regulate or give advice on the maximum gradients permissible for standard aerodromes. This maximum permissible gradient is generally in the order of 2%. However, when using private landing strips these maximum permissible gradients may exceed those which are normally accepted and particular care must be used.

Since the runway gradient has a direct bearing on the component weight of the aircraft, a runway gradient of 1% would provide a force component along the path of the aircraft which is 1% of the gross weight. In the case of an upslope, the additional drag and rolling friction caused by a 1% upslope can result in a 2% to 4% increase in the take-off distance and subsequent climb.

Frequently, the only runway at an airport has a slope. When determining which direction to use for take-off, pilots must remember that a direction uphill, but into a headwind, is generally preferred to a downwind take-off on a downsloping runway. Factors such as a steep slope, light wind, etc., may, however, make an uphill take-off impractical.

It is difficult to predict the retarding effect on the take-off run that water, snow/slush, sand, gravel, mud, or long grass on a runway will have, but these factors can be critical to the success of a take-off. Since the take-off data in the aircraft manual is predicted on a dry, hard surface, each pilot must develop individual guidelines for operations from other types of surface.

A typical general aviation aircraft manual states, 'for operation off a dry, grass runway, increase distance (both "ground run" and "total to clear 50 ft obstacle") by 7% of the "total to clear 50 ft obstacle" figures'. Grass, sand, mud, or deep snow can easily double the take-off distances. The pilot is responsible for determining this effect in the light of existing conditions.

Take-offs during cold weather

Take-offs in cold weather offer some distinct advantages but they also offer some special problems. A few points to remember are as follows:

- Do not over-boost supercharged or turbine engines. Use the applicable power charts for the pressure altitude and ambient temperature to determine the appropriate manifold pressure or engine pressure ratio. Care should be exercised in operating normally aspirated engines. Power output increases at about 1% for each ten degrees (F) of temperature below that of standard air.
- On multi-engine aircraft it must be remembered that the critical engine-out minimum control speed (Vmc) was determined at sea-level with a standard day temperature. If the appropriate power charts are not available to limit maximum rated power for take-off, Vmc will be higher than that published at below sea-level density altitudes.
- If icing conditions exist, use the anti-ice and de-ice equipment as outlined in the aircraft flight manual. If the aircraft is turbine powered, use the appropriate power charts for the condition bearing in mind that the use of bleed air will in most cases change the maximum load to be carried and the runway requirements.

Preflight preparation is the foundation of safe flying. Accident statistics of recent years indicate that adequate preflight preparation is lacking in many cases. Statistics indicate that many take-off accidents occur because:

- elements of the preflight preparation were not assigned the proper importance
- elements of the preflight preparation were not incorporated into the preflight routine
- pilots did not anticipate potential take-off emergencies and the required procedures to follow.

Although a pilot who is competent at instrument flying and radio navigation will undoubtedly be able to operate safely on more occasions than one with only a VMC capability, he will nevertheless from time to time still be faced with 'no-go' situations. In summary the combination of factors affecting the safety of a radio navigation flight comprise:

(1) The actual weather conditions at the time and those forecast for the period of the flight in relation to severity, e.g. thunderstorm activity, icing, etc. and the anticipated cloud base en route and at destinations.
(2) The equipment carried in the aircraft in relation to the ground aids available en route, at destination, and alternative.
(3) The need to have alternative navaid facilities available in case the facilities which are planned to be used, either on the ground or in the aircraft, become unserviceable during the flight.
(4) The pilot's instrument weather flying capability.

With regard to (4) above, it must be stressed that regardless of the degree of instrument flying competence achieved through training, this ability can only be sustained by current practice. Therefore, a pilot who is out of practice in instrument flying will need to keep this in mind during the flight planning stage and if it becomes apparent that instrument weather conditions prevail during the intended flight the pilot must consider the degree of weather to be experienced in relation to his current level of competence and **not hesitate to declare a 'no-go' situation if he feels that the weather situation is beyond his personal limits**.

Reducing the risk of inadequate preflight preparation

Some important actions which will enhance your safety and reduce the occurrence of unnecessary risks brought about by lack of or careless preflight preparation are:

- Form good preflight planning habits and review them continually.
- Be thoroughly knowledgeable of the hazards and conditions which would represent potential dangers, particularly take-off.
- Thoroughly familiarise yourself with your aircraft and engine handbooks in order to know intimately all systems, limitations and recommended operating procedures.
- Conduct your preflight planning and flight preparation with thoroughness to detail and appropriate to the intended operation.
- If you intend to use an aircraft as a means of transport for an appointment, always give yourself an out by informing your contact that you intend to fly and will arrive at a certain time, unless the weather conditions are unfavourable.
- Remember that a VFR pilot should avoid taking chances if the weather is marginal. Stay on the ground! A marginal weather operation in the winter is doubly hazardous since a pilot may be severely handicapped in selecting either an alternative course of action or a change in destination.
- Study the trend of the weather religiously in order to operate with maximum safety. Check all available weather information.

PREFLIGHT PLANNING AND PREPARATION

- Never fly into snow or rain showers which obscure the terrain. Use your good judgement and the 180° turn before you lose forward visibility and become a statistic.

Modern industry's record in providing reliable equipment is very good. When the pilot enters the aircraft, he becomes an integral part of the man–machine system. He is just as essential to a successful flight as the control surfaces. To ignore the pilot in preflight planning would be as senseless as failing to inspect the integrity of the control surfaces or any other vital part of the machine. Therefore before you enter the aircraft, assess your own fitness to fly.

During winter

- Remember that winter daylight hours are few and plan your flight accordingly. If your night experience is limited, be aware that night operation in winter can impose additional hazards.
- Remember to thoroughly warm the engine, cockpit and windscreen prior to take-off.
- Know that winter's low temperatures can change the viscosity of engine oils, reduce the effectiveness of the storage battery, and precipitate malfunctions in various component parts of your aircraft with little or no warning.
- Never attempt to take off with frost, ice or snow on the windshield, or on the wings and control surfaces of your aircraft.
- Have the following items checked for winter operation: cabin heater system for operation and leaks (**carbon monoxide can be deadly**), exhaust systems, windshield defrosting system, engine idle speed, carburettor heat, brakes, etc.
- Be forewarned that many pilots have inadvertently been placed on instruments, following a take-off in beautiful VFR weather, in aircraft that had been parked outside overnight. The condensation of moisture in the heater ducting completely covered the windshield from the inside. When conducting such an operation, make sure that the heater and air vents have been purged of moist air prior to take-off.
- Remember that during let-down, it may be difficult to keep the engine warm enough for high power operation, if needed. It may be desirable to use more power than normal during approaches to avoid excessive engine cooling. Remember that a rapid throttle operation with a cool engine may result in power loss or a hesitant response.
- Be alert during winter months for white-out conditions. Due to snow-covered terrain, haze, and falling snow, you could find yourself on instrument conditions with a complete loss of visual contact.
- Remember that depth perception is faulty when attempting to land on unbroken snow-covered surface or at night in marginal weather conditions.

- Remember that *you*, the pilot, have complete responsibility for the *go, no-go* decision based on the best information available – *do not* let compulsion take the place of good judgement.

Fig. 3.3 Don't gamble with the weather. (Photograph courtesy of UK Accident Investigation Branch)

To set off without adequate flight planning is akin to putting your life in your wallet and taking it to the nearest bookmaker.

4
Take-off and landing – performance aspects

In order for an aircraft to take off and land safely certain minimum distances must be available, and no pilot would disagree with this statement. Yet past accident figures show that every year in the UK over 10% of the total reportable accidents occur because pilots attempt to take off, or land, or go round again when insufficient distance is available to do so with safety.

Apart from these reportable occurrences there must also have been a significant number of incidents in which the damage sustained was fortunately of a minor nature and not sufficient to meet the definition of a reportable accident. In this respect most pilots will appreciate that any performance-related accident or incident could be classified as one in which a strong possibility of injury or worse could occur to the aircraft occupants, and therefore there is a clear need for all pilots to understand and guard against the underlying reasons for their occurrence.

So why do they occur? In all probability it boils down to four basic reasons:

(1) The take-off and landing performance for any specific type of aircraft varies widely and depends upon a number of different factors.
(2) During training any practical consideration of these factors is largely programmed out due to the size of the aerodrome being used relative to a light aircraft's performance.
(3) The assumption by a number of pilots that employing the procedure known as the short field take-off or landing technique will automatically resolve the problem.
(4) Human behaviour in the form of 'compulsion'; in this case it can be defined as an irresistible impulse to act in a certain way, against one's better judgement.

The exercise of good judgement in aviation can only be based upon knowledge, so keeping the previous facts in mind will give you a better opportunity to exercise good judgement during the planning of any flight and go a long way to avoiding your inclusion in the accident statistics. So let's take a closer look at the four basic reasons put forward above.

Factors affecting take-off performance

There is little point in knowing the length of a runway unless you can relate this to the anticipated performance of the aircraft in the conditions which exist at the time. These conditions could be simply summed up by the following question:

Q. How long is a runway?

Is the question too vague for an answer? Not really, because it can be answered very specifically as follows:

A. As long as will be allowed by the following factors:
   The type of aircraft using it.
   The wind velocity.
   The weight of the aircraft.
   The air density (pressure altitude and temperature).
   The type of surface.

Thus the prudent pilot will need to give consideration to performance calculations before doing any other form of flight planning or he could find, at best, that he is wasting his time contemplating the particular flight, or at worst, programming himself into a no-go situation which he will most likely discover too late.

Table 4.1 Take-off performance

Example: Take-off run in zero wind, sea-level, paved, level, dry runway and ISA conditions ... 800 feet

Factors affecting take-off performance		Increased take-off run (feet)
Wind	A tailwind of 10% of the take-off speed will add 20% to the length of the take-off run.	960
Weight	An increase of 10% over the all-up maximum weight will add 5% to the lift-off speed and 20% to the take-off run.	1150
Altitude	A take-off at 1000 amsl will add 10% to the take-off run.	1265
Temperature	A temperature of 10°C above ISA will add 10% to the take-off run.	1392
Surface	An upslope of 2% will add a minimum of 10% and often more to the take-off run.	
	Soft ground will increase the take-off run by 20% at least.	2204
	Long grass will increase the take-off run by 20% at least.	

In order to put the various factors affecting take-off performance into perspective and see how they combine to alter the aircraft's ability to take off within a given distance, an example is given in Table 4.1, which shows how the various factors can combine to lengthen the required take-off run.

Due to the variations incurred by such items as thrust:weight ratio etc., it is not possible to give exact percentages applicable to all light aircraft. Therefore the percentages given in the table are only approximate, but nevertheless give a good working guide when assessing how the different factors affect an aircraft's performance.

The figures in Table 4.1 are cumulative and if all these conditions are present, then the final figure of 2,204 feet will be approximately correct. Thus we see that a typical light aircraft which is capable of becoming airborne in 800 feet under ISA conditions at sea-level in zero wind will take nearly three times longer when the specific effects of all the above factors are applied. In addition the effect of the weight, altitude and temperature figures shown in the table would reduce the rate of climb by about 40%, a very significant factor if obstructions lie ahead of the take-off path.

Wind

In considering the effect of wind as shown in Table 4.1, it should be borne in mind that a typical lift-off speed for a light aircraft is in the region of 50 knots, so a tailwind component of 10% is equal to a tailwind of only 5 knots and the visual effect of a wind of this light strength will give a very small indication from a windsock. Additionally it should also be appreciated that most take-offs are carried out into a headwind of at least 10 knots or more, thus the average take-off run experienced by pilots will normally be some 40% shorter than that experienced in zero wind conditions, and therefore a 5-knot tailwind could produce a take off run which is 60% longer than that which a pilot is customarily used to.

Weight

As far as a 10% increase in weight is concerned, this could easily be produced by loading a couple of heavy suitcases or topping up the fuel tanks of a four seat aircraft with four people on board.

Altitude/Temperature

The altitude factor may be completely ignored by pilots who have been trained in those countries where most aerodromes are only a little higher than sea-level. This lack of awareness could cause serious problems when operating in those countries where high elevation aerodromes are common;

for example, over 70 aerodromes in France are over 1,000 amsl. Again, pilots trained in the UK for example do not often come up against high temperatures but if they venture into Europe they will find the situation very different in that Southern Europe during the summer months has a mean temperature of some 30°C during the day, excellent for taking off clothes but not so good for taking off an aircraft.

Another sneaky effect is that of water vapour or high humidity, which can affect engine power output significantly, and this should be taken into consideration when planning take-offs in muggy or high humidity conditions. Whenever water vapour is present there is less air entering the engine. Secondly, this creates an excessively rich mixture because the amount of fuel is the same but the density of the air is less, and finally the water vapour slows the rate of fuel burning. These factors will all bring about a loss of power. Therefore when warm humid air is present it will be advisable to add between 10% and 15% to the figure calculated for take-off run and take-off distance.

Surface conditions

With regard to the possible variations in surface conditions it is not very easy to assess with accuracy the retarding influence of different surfaces. How does one measure surface gradient, softness of the ground, or the length of the grass in relation to its anticipated effect on the length of the take-off run? One has only to try to push a car uphill or along soft ground or through long grass to appreciate what a significant effect these factors can have upon an aircraft's performance. However, one thing is certain: if the retarding effect of the ground is such that the aircraft cannot accelerate to its lift-off speed then no take-off will be possible, regardless of the length of take-off run available.

Incidentally, one way of determining the degree of slope along a landing strip when the runway length is known with reasonable accuracy, is to taxy the full length of the runway and note the altimeter reading at each end (most altimeters are calibrated in 20-ft divisions). Then divide the altitude difference between each end by the length of the landing strip, e.g. an altitude difference of 25 feet on a 2,500 ft strip will indicate a 1% slope. However, when runway lengths are given in metres be sure to convert these to feet before your calculation.

So much for the dangers of ignoring the need for performance calculations – but what of the advice that can be given in this respect? Well, simply stated, it will not always be necessary for calculations of this nature to be conducted before take-off but whenever conditions indicate that any of the factors discussed might have an adverse affect upon the aircraft's performance the pilot must refer to the aircraft manual and aerodrome data both for take-off and landing destination to ensure that he is able to conduct the flight with safety. When carrying out these calculations it would be

TAKE-OFF AND LANDING -- PERFORMANCE ASPECTS

advisable to add a further 20% to the run/distance required figure just to ensure that the effect of any wind change during the take-off or landing, or any other variation of the factors involved, are also taken into account.

TAKE-OFF FACTORS		Feet
Example - ISA. Zero Wind. Sea Level.		800
WIND	Tailwind 10% of take-off speed	960
WEIGHT	10% over maximum permitted	1150
ALTITUDE	1000 feet a.m.s.l.	1265
TEMPERATURE	25 degrees C.	1520
SURFACE	Upslope 2% Soft ground Long grass	2640

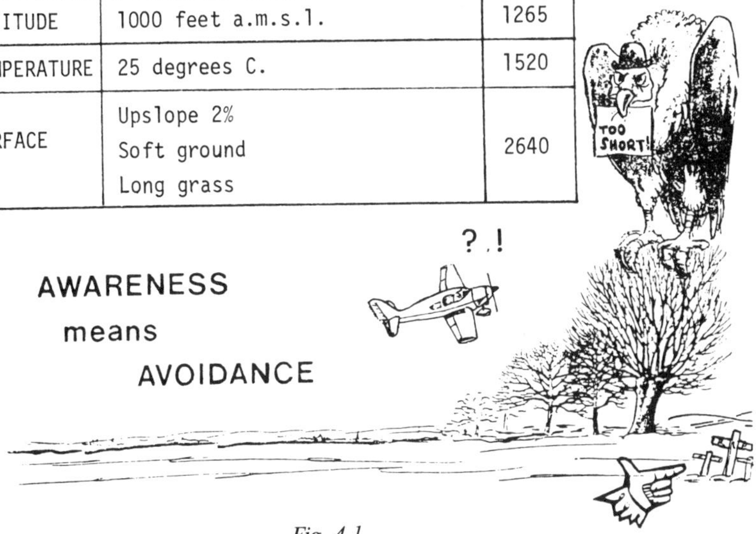

AWARENESS

means

AVOIDANCE

Fig. 4.1

So far, the factors affecting take-off performance have been considered in broad detail without reference to a specific aircraft. It would therefore be of interest to take a typical light aircraft in current use today and relate the various factors previously discussed to the relevant performance information given in its aircraft manual.

We will assume that the pilot has given little consideration to the factors which will affect the performance of the aircraft during the take-off and has loaded it in a manner that produces the following figures:

(a) The aircraft is equipped with long range tanks which are full: 40 imp gal, or
    48 US gal   288 lb
(b) 1 pilot and 3 passengers occupy the 4 seats: 720 lb
(c) Baggage (weekend away for 4 people), say: 160 lb
(d) Basic weight of the aircraft (with its fitted
    equipment):   1477 lb
                                               Total weight: 2645 lb

TAKE-OFF DATA

Take-off distance from hard surface runway with flaps up

Gross weight (lb)	IAS at 50' (knots)	Head wind (knots)	At sea-level and 59° Ground run	At sea-level and 59° Total to clear 50 ft OBS	At 2,500 ft and 50°F Ground run	At 2,500 ft and 50°F Total to clear 50 ft OBS	At 5,000 ft and 41°F Ground run	At 5,000 ft and 41°F Total to clear 50 ft OBS	At 7,500 ft and 32°F Ground run	At 7,500 ft and 32°F Total to clear 50 ft OBS
2300	58	0	865	1525	1040	1910	1255	2480	1565	3855
		10	615	1170	750	1485	920	1955	1160	3110
		20	405	850	505	1100	630	1480	810	2425
2000	53	0	630	1095	755	1325	905	1625	1120	2155
		10	435	820	530	1005	645	1250	810	1685
		20	275	580	340	720	425	910	595	1255
1700	48	0	435	780	520	920	625	1095	765	1370
		10	290	570	355	680	430	820	535	1040
		20	175	385	215	470	270	575	345	745

Notes:
(1) Increase distance 10% for each 25°F above standard temperature for particular altitude.
(2) For operation on a dry, grass runway, increase distance (both 'ground run' and 'total to clear 50 ft obstacle') by 7% of the 'total to clear 50 ft obstacle' figure.

MAXIMUM RATE-OF-CLIMB DATA

Gross weight (lb)	At sea-level and 59°F IAS (knots)	At sea-level and 59°F Rate of climb (ft/min)	At sea-level and 59°F Gal. of fuel used	At 5,000 ft and 41°F IAS (knots)	At 5,000 ft and 41°F Rate of climb (ft/min)	At 5,000 ft and 41°F From S.L. fuel used	At 10,000 ft and 23°F IAS (knots)	At 10,000 ft and 23°F Rate of climb (ft/min)	At 10,000 ft and 23°F From S.L. fuel used	At 15,000 ft and 5°F IAS (knots)	At 15,000 ft and 5°F Rate of climb (ft/min)	At 15,000 ft and 5°F From S.L. fuel used
2300	79	645	1.0	74	435	2.6	70	230	4.8	65	20	11.5
2000	76	840	1.0	70	610	2.2	65	380	3.6	59	155	6.3
1700	72	1085	1.0	67	825	1.9	61	570	2.9	56	315	4.4

Notes:
(1) Flaps up, full throttle, mixture leaned for smooth operation above 3,000 ft.
(2) Fuel used includes warm up and take-off allowance.
(3) For hot weather, decrease rate of climb 20 ft/min for each 10°F above standard day temperature for particular altitude.

Fig. 4.2

The aircraft is 345 lb overweight; this is equal to 15% over the limit for the maximum allowable weight for the aircraft. 15% over the maximum permitted weight will, apart from exceeding the manufacturer's weight limitation, increase the take-off run by 30%.

The aircraft take-off data as shown in Fig. 4.2 reveals that under ISA, sea-level, zero wind conditions and a maximum gross weight of 2,300 lb the take-off run will be 865 feet. Allowing for the added weight this will be increased by 30% ... **1124 feet**

TAKE-OFF AND LANDING – PERFORMANCE ASPECTS

Assuming the aircraft is being taken off from a site which is 1,000 feet above sea-level, then interpolating the take-off data in Fig. 4.2 we have to add at least 70 feet (nearly 10% of the take-off run at sea level) ...

1194 feet

Let us now say it is summer and the temperature at the planned time of take-off is 28°C. We see in note (1) in Fig. 4.2 that the manufacturers are stating that a 10% increase in the take-off figures should be allowed for every 25°F (13°C) above ISA. In this case the take-off run now becomes ...

1314 feet

We now come to the effect of wind. Assume in this case a 5-knot tailwind component exists. We know from the information given in Fig. 4.2 that a headwind of 10 knots reduces the take-off run by 250 feet, so applying half this in reverse to equate to the tailwind of 5 knots we have an increase in the take-off run of 125 feet, i.e. approximately 15%.

$$15\% \text{ of } 1314 = 197$$
$$197 + 1314 = \mathbf{1511 \text{ feet}}$$

If the take-off is being carried out from a dry grass surface this figure must be increased by 7% (see note (2) in Fig. 4.2).

$$7\% \text{ of } 1511 = 106$$
$$106 + 1511 = \mathbf{1617 \text{ feet}}$$

If, however, the grass surface is wet, an additional factor must be allowed, and this should be at least 20%, so the take-off run figure will now be ...

1940 feet

Finally if a small upslope is present along the take-off path and the grass is not short then another 30% should be added for safety, and this will bring the take-off run to ...

2522 feet

From these calculations it will be seen that the actual take-off run will probably be some three times longer than quoted in the first line of the take-off data given in Fig. 4.2. This figure could be longer than that available at some normal aerodromes.

These figures so far have been based upon the fact that the aircraft is in a reasonable condition and the power plant is 100% efficient; they do not allow for aircraft age or any slight deficiency in power available so it would be wise to add a further 10% to allow for any deterioration in aircraft performance. Thus the final figure compatible with safety would be ...

2775 feet

The story however does not end here because we have to consider the take-off distance, i.e. the distance from the commencement of the take-off run to clearing a 50-foot obstacle. In the circumstances already described it should be borne in mind that the aircraft's rate of climb could be decreased by 40%. Allowing for the fact that the take-off distance will be affected by the length of the take-off run and then allowing for the reduction in the aircraft's climb performance it can be appreciated that the take-off distance

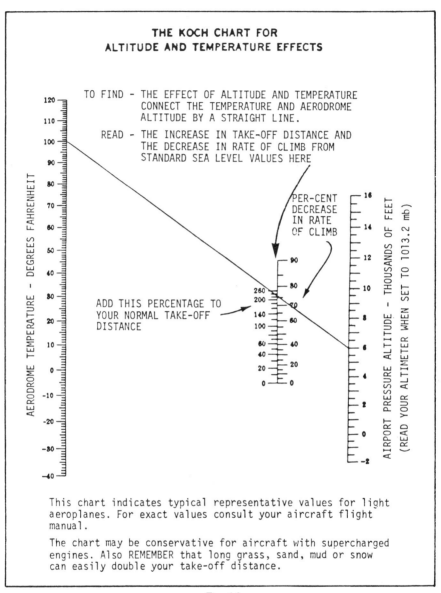

Fig. 4.3

in this case will be between 3,500 and 4,000 feet. Bearing all these points in mind it will be obvious to any sensible person that a pilot who just lines up and takes off without giving any consideration to the performance factors will be operating in an area of extreme risk.

Before leaving the subject of take-off performance it should be pointed out that a situation may apply where the pilot is flying an older type of aircraft for which there is no aircraft manual or take-off performance

TAKE-OFF AND LANDING – PERFORMANCE ASPECTS

information. In these circumstances reference to a Koch Chart (Fig. 4.3) will give a good approximation of the effects of altitude and temperature during take-off.

Example: In Fig. 4.3 the diagonal line shows that 230% must be added for a temperature of 100°F and a pressure altitude of 6,000 feet. Therefore if the standard temperature sea-level distance in order to climb to 50 feet normally requires 1,000 feet of runway, it would become 3,300 feet under the conditions shown. In addition the rate of climb would be decreased by 76%. Thus, if the normal sea-level rate of climb for the aircraft is 500 feet per minute, it would become 120 feet per minute.

Factors affecting landing performance

Similar factors affect an aeroplane during the landing phase and although the figures shown in Table 4.2 are approximate they are based upon a typical light aircraft and give a fair idea of how landing distance can be increased through the adverse effects of these factors.

A further point in considering the effects of the various take-off and landing performance factors is that on reviewing past causes of this type of accident it becomes evident that in most cases it was not just ignoring one of these factors which produced the accident, but rather the combined effect of several factors which, taken individually, were not very large.

Table 4.2 Landing performance

Example: Landing distance in zero wind, sea-level, ISA conditions, flaps 30°, airspeed at 50 feet 55 knots ... 1,200 feet		
	Factors affecting landing performance	Increased landing distances (feet)
Wind	A tailwind of 10% of the touchdown speed will increase the landing distance by 20%	1440
Weight	An increase of 10% above the maximum all-up weight can add 20% to the landing distance.	1728
Altitude	A landing at an aerodrome 1,000 feet amsl can add a minimum of 10% to the landing distance.	1900
Temperature	A temperature of 10°C above ISA can add a minimum of 5% to the landing distance.	1995
	A downslope of 2% will add a minimum of 10% and often more to the landing run.	2195
Surface	A wet or slippery surface can add up to 50% or more to the landing run because of reduced retardation and braking efficiency.	3292 (Cumulative total)

The size of the training aerodrome

During his training a pilot will normally be operating from aerodromes with runways or take-off and landing areas which are long enough to provide an excess of the distance required for most light aircraft, even on those occasions when the various performance factors are adverse. This often leads to a lack of awareness of the correct lift-off speeds and the correct application of proper approach speeds. Whereas the use of proper approach speeds in relation to landing performance is fairly easy to understand, the use of correct lift-off speeds is not always appreciated, so a few words on this subject would be very pertinent for the reader. There are two basic reasons for knowing the correct lift-off speed; one concerns the fact that the figures given in the aircraft manual and which relate to take-off run and take-off distance are based upon a specific lift-off speed. This airspeed normally equates to 1.15 or 1.2 times the stall speed, V_{si} in the case of a flapless take-off or V_{so} in the case of a take-off using flap.

Even allowing for the aircraft's acceleration, any attempt to lift the aircraft into the air at the V speed compatible with its weight could produce a hazardous situation where the pilot only has marginal control. However, because of the rapid acceleration at this point only a small increment needs to be added to the stalling speed to produce the necessary safety factor.

If the pilot allows the aircraft to remain on the ground beyond the recommended lift-off speed the take-off run and take-off distance will naturally be increased and the performance figures specified for the aircraft will thus be degraded. All very obvious, but what is not so clearly appreciated is the degree of effect this has on the established performance figures. For example, the take-off distance varies as the square of the take-off velocity, thus a lift-off speed 10% above that specified will result in a 20% increase to the take-off distance. In the case of a light aircraft which has a published lift-off speed of 50 knots it can be seen that a small addition of 5 knots before lifting off will increase the take-off distance by approximately 20% and this is a very substantial increase.

Thus it becomes clear that just applying a little back pressure to the control column during the take-off roll and allowing the aircraft to become airborne when it feels like it is a very crude method of taking off. You can note this for yourself the next time you fly, just apply a little back pressure and note the speed at which the aircraft elects to become airborne. In a number of typical light aircraft used today (and depending upon the amount of back pressure used) you will be surprised to see how unnecessarily high the airspeed becomes at the point of getting airborne when using this method. One reason for pilots tending to attach insufficient importance to accurate lift-off speeds is that many aircraft manuals do not include this figure in the 'normal operating procedures' section of the manual. Nevertheless, the figure is usually quoted in the graphs or tables which refer to 'take-off performance'.

Another reason for the lack of emphasis on accurate lift-off speeds during training is no doubt a hangover from earlier days in that in tail wheel aircraft the greatest difficulty during the early stages of training in take-offs was in keeping the aircraft on a straight path and any distraction from looking ahead could have caused problems in directional control. However, with the advent of the nose wheel aircraft this has in the main been eliminated, so enabling the pilot to monitor more easily and safely such items as airspeed and engine instruments during the take-off roll; but more of that later.

Returning to general performance considerations, it is a fact that student pilots who rarely have a need to carry out performance calculations during training will also have a strong tendency to exhibit a lack of awareness in this field later on in their flying activities, and this will be particularly hazardous when operating from small landing strips. In many countries today, the number of private strips exceeds the number of normal aerodromes and, bearing in mind the escalating costs involved in flying, many pilots are quite naturally attracted to using these strips more and more. The risk element in ignoring take-off and landing performance considerations is therefore becoming higher.

The problems of the short take-off technique

A few words of caution must be written concerning the short field take-off technique. It is taught during training for a pilot's licence but its practical use is extremely limited in that, although it is related to obtaining a better obstacle clearance when the take-off distance available is rather short, the automatic assumption that it will resolve a short field situation can lead to disaster.

One has only to consider the situation in which it will most likely be used, for example a take-off from a landing strip or farm field. In these circumstances how does one assess the length of the available ground run with any accuracy? How does one determine the strength and direction of the wind relative to the intended take-off path? How does one measure the height of any obstructions? After all, pilots are not in the habit of carrying with them hand-held anemometers or theodolites. Again, how does one measure a 1% or 2% upslope or downslope against the irregular terrain surrounding the field being used? What about altitude, temperature, etc.? Clearly, if a tight situation is felt to exist then all the normal performance calculations will be vital and they will need to be done very accurately, but it is in these very circumstances that the calculations (if made at all) will usually be inspired guesswork.

When facts and figures are not readily available to make sensible performance calculations, then bear in mind a rule of thumb known as 'the half-way rule'. This is based upon the fact that if an aircraft has not reached

Fig. 4.4

its lift-off speed at the half-way point you are moving into a high risk situation and your most sensible action would be to abort the take-off. The reason for this is that when light aircraft operate from aerodromes or private landing strips of reasonable length it will be seldom that the the take-off run will exceed more than half the length of the runway or take-off area. Thus, if the lift-off speed has not been achieved by the half-way point it is indicative of a potentially dangerous situation which may be the result of one or more adverse factors, some of which are given below:

(1) The engine may not be developing full power.
(2) The aircraft brakes may be binding, or being inadvertently applied.
(3) The surface is softer than anticipated (and may get softer as you continue).
(4) The aircraft is overweight, or in any event too heavy to take off in the length available.

TAKE-OFF AND LANDING – PERFORMANCE ASPECTS

(5) A tailwind component is being experienced.

The use of the 'half-way rule' can be a simple and effective aid to a pilot's decision-making and reduce the risk of continuing with a difficult or impossible take-off situation.

To attempt to take off in the belief that the employment of the short field technique will by itself provide sufficient safety would be foolhardy and more aptly described as a '**leap into the unknown**'. However, these comments are not intended to imply that the short take-off technique should be avoided as an operational procedure, but rather its use should be confined to those situations where it can be ensured that sufficient take-off distance is available but where an added safety factor would be an advantage.

It can generally be stated that the use of flaps during take-off will permit a reduction in the length of the take-off run and cause a small deterioration in rate of climb. The aircraft manual should be consulted to see whether and under what take-off conditions the use of flaps is recommended.

For many years the use of flaps has been associated with the obstacle clearance take-off but their effectiveness in achieving a better obstacle clearance is shrouded more with opinions than with facts, in that manufacturers rarely give detailed performance figures for take-off distance with the flaps in use.

However, one value in the use of flaps during take-off is clear and that is their effect upon the aircraft's stalling speed. If the stalling speed is lowered by, say, 5 knots then the lift-off speed can be achieved earlier and in light aircraft this means a shorter take-off run as the increased drag from a small increment of flap is extremely nominal due to the low lift-off speed.

A second value, i.e. a lower 'climb out' speed, is however not always as beneficial as it seems. During the short take-off procedure it is normally recommended that the aircraft is initially flown at Vx (best angle of climb speed) following lift-off. This is on the premise that the aircraft, in flying a little more slowly, will take longer to reach any obstacles and thus permit it to achieve a greater height above such obstacles.

However, to keep this small safety factor in perspective some figures, based upon simple arithmetic, are given in Table 4.3. These give some comparisons in relation to height achieved for distance travelled by an aircraft flying at 70 mph and 60 mph respectively. The height achieved in each case is based upon a 500 fpm rate of climb. It should nevertheless be appreciated that the lower airspeed, particularly if flap is used, will, most likely, produce a reduced rate of climb.

The dangers of compulsive actions

Finally we come to compulsion, a trait which appears so commonly as a cause factor in accidents; if only pilots traded compulsion for common sense the bulk of the accidents would disappear. The reasons for the display of

Table 4.3

Speed (mph)	Distance (feet)	Time (seconds)	Height (feet)	Vertical differences (feet)
70	800	7.79	64.93	10.82
60		9.09	75.75	
70	1000	9.74	81.16	13.53
60		11.30	94.69	
70	1200	11.68	97.40	16.23
60		13.63	113.63	
70	1320	12.85	107.14	17.86
60	(¼ mile)	15.00	125.00	

Climb rate: 500 fpm at 70 mph (best rate of climb speed)

compulsion in human beings are more a matter for the psychologist to explain than for discussion in this book. What can be said here is that all pilots who fall into the compulsion trap are accidents waiting to happen, and there is no substitute for common sense based upon knowledge, self-discipline and good habits when flying an aircraft. One has only to review the number of landing accidents which occur each year to see that few would have been attributed to pilot error if the pilot had only used his right arm properly – in this case to open the throttle and go round again was all that was needed to prevent a hazardous occurrence or a bent aeroplane, and it cannot be stated more simply than that.

In defence of the pilot, and before leaving the subject of compulsion, it will be pertinent to refer to how habit training can sometimes be on the side of bad judgement. It is an acknowledged fact that when emergency situations occur pilot reaction will, in the first instance, usually be controlled by habits, either those taught during training for specific situations, or those picked up during past flying experience. How these habits can sometimes be the cause of accidents is mentioned at various times throughout this book but one illustration regarding subconscious habits, which is applicable to the take-off situation is as follows.

Many habits manifest themselves in physical actions; for example, during the process of taking off the pilot opens the throttle(s) fully and this is a habitual action taken following a situation which is heavily underscored by the anticipation of conducting a take-off from the moment he commenced to taxy. Now, assume that during the take-off run the pilot on checking the engine instruments notices that the oil pressure gauge is reading zero. During this check and the immediate decision which has to be made, there will be a strong tendency for habit to create an instinctive resistance to, and sometimes overwhelm, a pilot's common sense which will be telling him to close the throttle and discontinue. What is not always appreciated in this type of situation where time is at a premium, is that the pilot made his original decision when he first opened the throttle, a decision which had been made many times before an in each case there had been no subsequent problem. Alas, on this occasion a very real problem occurs and now his thought processes have to be jerked away from whatever else he is thinking and applied to the business of making an unexpected decision, at a time when other factors such as controlling the aircraft, assessing the length of runway remaining, etc., will be requiring his attention. Throughout this stage the strong habit action of keeping the throttle fully open will be present and may be strong enough merely to cause a delay – and so prevent the correct decision and follow-up action being made before the aircraft becomes airborne. If in these circumstances the oil pressure indications are correct an engine failure after take-off will be inevitable. (In this connection it is worthy of note that in the UK an average of 10 accidents occur each year due to the engine failing within a minute or so of taking off. There are of course incidents which are not included in the accident statistics because the pilot was fortunate enough to land the aircraft without damage.)

The best way of overcoming habit problems of this sort will be for training and practice to include simulated emergencies of this nature whereby the pilot carries out an aborted take-off from time to time.

Finally, to appreciate the importance of how lack of attention to the various performance factors adds considerably to the risks involved during take-offs and landings we only have to read the following extracts taken from the accident summaries and which are but a few of the typical performance accidents which occur each year ... the message is clear!

> 'On a private strip of just over 300 metres the pilot opened the throttle fully against the brakes, prior to take-off the aircraft rotated but the landing gear struck the stone boundary wall, following which the aircraft nosed over into the adjoining field ...'
>
> 'The pilot carried out a landing on wet grass using a short field technique but the aircraft failed to respond to braking action. The nose wheel broke off in the overrun ...'
>
> 'The pilot took off from a farm strip but failed to accelerate beyond rotation speed. The pilot managed to get the aircraft airborne but the wheels came into contact with the boundary edge. The aircraft landed in the next field, breaking off the landing gear in soft ground ...'
>
> 'The touchdown was made half-way along the runway but braking action was poor on the wet grass. The aircraft overran into a ditch ...'
>
> 'Shortly after taking off from a farm strip the pilot made a distress call stating "structural damage". The aircraft landed nose down into trees, all persons on board were killed ...'
>
> 'During a take-off from a field the aircraft failed to clear the boundary hedge ...'
>
> 'The aircraft overran the landing strip and collided with a fence ...'
>
> 'During take-off the pilot was not satisfied with the acceleration

and abandoned the take-off. The aircraft overran into a hedge. The short strip had a soft surface and the grass was longer and thicker than the pilot expected ...'

'The pilot aborted the take-off when the aircraft showed no inclination to become airborne. Braking action proved ineffective and the aircraft struck the boundary hedge ...'

'After reaching rotation speed the aircraft failed to climb away and crash-landed in the overrun area ...'

'During the take-off run the pilot realised that the aircraft was very slow to accelerate. Half-way along the runway, although it was not possible to take off within the boundary, the pilot elected to continue. The aircraft failed to become airborne and after traversing three hedges, a field and a road, came to rest in a second field. The pilot left the aircraft via the port door but as he did so the fuel caught fire and the violent explosion blew him off the wing. The door slammed shut and jammed. The front occupant eventually kicked it free and the rear passengers kicked out the port rear window and exited through it. Subsequent examination revealed all three tyres were well below their recommended pressures. The mean upslope of the grass runway was 2.0%. With the grass length and high humidity as additional factors the take-off performance would have been further degraded ...'

'The pilot made a go-around from his first approach having arrived at the airfield in patchy stratus. The second approach resulted in the aircraft touching down some two-thirds down the runway. Heavy braking on short wet grass caused the aircraft to skid along the remaining runway and it slid off the end, hitting a gate and fence ...'

'The pilot had visited the strip by road to examine it from the ground before flying in with another pilot as passenger to fly out another aircraft. The pilot flew low over the airstrip and then made another approach. It touched down well inside the boundary, moving fast and skipping across the surface. At a late stage, power was applied and the aircraft became airborne in a steep nose-high attitude. The aircraft passed through the tops of some trees and then sank, hitting power lines before crashing into a sewage works ...'

Finally, the following extract from a pilot's report after colliding with a dry stone wall when attempting to take off from a small field should punch home the dangers of attempting a short field take-off without making a few performance calculations first.

'However, I did use the longest run and I did lower flaps as recommended in the aircraft manual, and I did open up to full power against the brakes just as I had been taught, but it didn't work ...'

It was indeed fortunate that on this occasion the words above did not become the pilot's epitaph because, referring to the field length available and the performance figures in the aircraft manual, it was clearly obvious that there was no possibility of the aircraft becoming airborne before hitting the stone wall, still less was there any chance of clearing the trees just beyond. In this case the pilot and passengers survived because the dry wall got in the way whilst the aircraft was still on the ground. If it had got airborne the inevitable collision with the trees would have resulted in several 'wills' being read.

Reducing the risk of a performance accident

Taking into account how the various factors affect the performance of an aircraft during take-off and landing it becomes very clear that your risks will be reduced by adopting the following safeguard actions:

- Wind. Whenever the wind is light or variable give consideration to the need to calculate take-off run and distance. The smaller the aerodrome or take-off strip, the greater the need.

  Whenever strong winds or strong crosswinds prevail give due consideration to the need for your flight to take place; such winds will give difficulty during take-off and more particularly during your approach and landing. Bear in mind the crosswind limitations as shown in your aircraft manual and remember the aircraft has not been demonstrated as being capable of safely taking off or landing in crosswind components greater than shown. You owe it to yourself and your passengers not to become an explorer of an aircraft's capabilities – leave this to the test pilot.
- Altitude/temperature. When operating from high altitude airfields or from any airfield when high temperatures prevail (particularly if heavily laden) a performance calculation should be considered a *must* before you fly.
- Humidity. Bear in mind that 'humidity' affects an aircraft's performance in an insidious manner, for example on those warm, muggy days if humidity reduces air density by 3% it will result in about a 12% reduction in the aircraft performance.
- Surface condition. An adverse surface condition during take-off, e.g. upslope, soft ground, long, grass, snow, etc., could result in a situation where the aircraft is incapable of reaching its lift-off speed. So make a suitable allowance and be prepared to abandon the take-off whilst sufficient stopping distance remains, i.e. if you have not become airborne at the half-way point, abort the take-off and reconsider the situation. This will be particularly important when taking off from small strips. During landing bear in mind that a downslope will add considerably to your stopping distance.
- Lift-off speed. Know the correct lift-off speed for the aircraft type and weight: this information, together with the correct decision, would have prevented many past accidents from occurring. Armed with this

knowledge many pilots involved in take-off accidents would have recognised, from the speed achieved and the distance remaining, that there was no possibility of becoming airborne safely and abort action would more likely have been taken, or taken more promptly.

> **Bear in mind that a review of past take-off and landing accidents proves that, in the majority of cases, it was the combined effects of several performance factors which caused the accidents, rather than one factor alone.**

5
The take-off – procedures and control

Apart from the apparent lack of knowledge and proper attention to the performance aspects, past take-off accident reports also reveal a number of items concerning the procedures and methods of controlling the aircraft in which carelessness or the use of incorrect procedures were the cause of incidents and accidents during the take-off and landing phases. The following comments relate to the most common of these and concern that period from the time the power checks are carried out to the initial climb following take-off.

Use of check-lists

Bearing in mind the advanced technology which is an integral part of today's aircraft and the items which make up the equipment and systems of the machine it is highly recommended that a 'check-list' should be used. A pilot's memory is not infallible, added to which is the fact that general aviation pilots quite commonly fly aircraft of different types with irregular frequency. Therefore common sense and caution both demand the use of some form of memory aid.

A check-list can be likened to a type of insurance policy but just as one should read the small print and understand the contents of such a policy so should pilots be careful to interpret the purpose of a check-list's contents and above all not treat it like a shopping list, i.e. merely to jog the memory.

Firstly, the check-list used must be the one applicable to the aircraft type and not one which is merely identified with the generic series of aircraft. For example, many popular light aircraft, e.g. the Cessna 172, have been manufactured over a large number of years and each year's model brings with it changes in equipment, operating speeds and limitations. Therefore the check-list used must be the one produced not merely for the particular type of aircraft but also for the correct model.

Remember too that any model of aircraft may have been modified by a previous owner and the check-list may need amending accordingly. More obvious instances of this would be any modification to the fuel system or the incorporation of a Robertson short field take-off and landing modification.

Some check-lists leave much to be desired in their format; a typical example and a trap for the unwary pilot is where two items have been incorporated on one line. This can easily lead to the pilot carrying out the first action itemised on the particular line of the check-list and then dropping his eyes down to the next line and continuing his checks, having omitted the second item on the previous line.

Finally, as indicated earlier, the check-list should not be read off like a piece of poetry but used to stimulate not only the memory, but also the pilot's thinking process in relation to why certain actions are being carried out and how they should be applied to the prevailing conditions.

During the power checks

The importance of choosing a suitable area free from loose stones and similar debris appears to be forgotten on sufficient occasions to cause a number of accidents each year due to propeller disintegration during flight.

Such incidents to single-engine aircraft usually result in a forced landing away from an aerodrome, and to expose oneself and others to such a situation with its high risk of subsequent injury is needlessly brought about by lack of care during the power checks and/or the initial stage of the take-off run. In the case of the latter, when loose stones or deteriorating runway surfaces are present at the commencement of the take-off run, the throttle(s) should be opened slowly to avoid debris being sucked up and damaging the propeller.

An additional factor in relation to the foregoing comments is the need to examine the propeller carefully, particularly the rear face, during the preflight inspection, and in this connection the following extract from a UK accident bulletin is reproduced below:

Propeller failure from stone damage

The photos [Figs. 5.1 and 5.2] show the result of stone damage to the propeller. The front face was newly polished and lacquered following overhaul, however the rear face contained a large number of surface blemishes, any one of which could have led to fatigue fracture (and did).

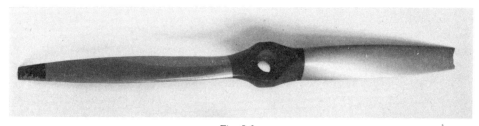

Fig. 5.1

THE TAKE-OFF – PROCEDURES AND CONTROL

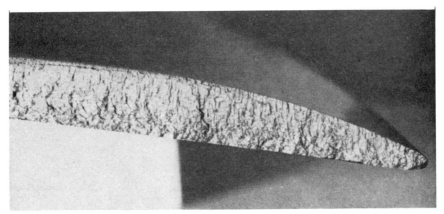

Fig. 5.2 Failed end section of propeller blade.

Pilots and engineers should note that it is the *rear face* of the propeller which is more critical than the more often inspected front face.'

In relation to propeller damage, it may be of some interest to note that the first fatal aeroplane accident occurred in 1908 when Orville Wright's passenger, Lieutenant Selfridge, was killed in a crash following propeller failure.

Magneto and rpm checks

Two items applicable to this aspect of the power checks and concerning risk factors are:

(1) The rpm drop when a magneto is switched off.
(2) The sequence in which a 'key ignition' switch is operated.

In the case of (1) above, the aircraft manual normally gives two figures concerning the allowable mag drop; one relates to the maximum drop permitted and the second relates to the difference between the 'mag drop' from each magneto. Most pilots are fully aware that a single mag drop greater than that shown in the aircraft manual means a no-go situation. However, the differential mag drop check is also equally important and if this figure is greater than given in the manual then this also indicates a 'no-go' situation.

The second item under this heading relates to the magneto selection sequence when operating aircraft with 'key start' systems. Some engine failures have occurred due to the malfunction of a single magneto because the ignition key was selected to one magneto only, by mistake. How this mistake may occur can be traced back to the power checks.

Many pilots carry out the magneto checks in the following sequence:

> 1st action – The ignition key is turned from the 'Both' position to the adjacent 'Left' position, then back to 'Both'.

Fig. 5.3

    2nd action – The ignition key is then turned from the 'Both' position through 'Left' to the 'Right' position and then back to 'Both'.

However, during this procedure most pilots are instinctively aware that if they overshoot when returning the key back to the 'Both' position they will activate the starter motor, causing damage. Thus during this operation there is an inherent possibility that the key may finish up in the 'Left' position instead of 'Both'. The scene is now set for a small reduction in power available during the take-off and if the 'Left' magneto should fail at any time during the flight the pilot and passengers will be exposed to a high risk situation.

    Whilst fully appreciating that pilots are taught to incorporate ignition checks during their pre-take off checks and also during their engine failure checks should an engine stop in flight, it can only be stressed that pressures and distractions may interfere with such procedures and if the relevant magneto fails just after take-off, there may not be sufficient time available to carry out this action. Therefore although a pilot cannot guarantee against making an ignition switch error it is suggested that the following sequence be used during the magneto checks prior to flight, as it does reduce the possibility of finishing up on one magneto.

    1st action – Turn the ignition key from the 'Both' position to the 'Right' (furthest) magneto and then back to 'Both'.
    2nd action – Turn the ignition key from the 'Both' position to the 'Left' (nearest) magneto – then one 'click' back to 'Both'.

    Whilst on the subject of ignition checks it would be appropriate to mention that the selection of carburettor heat (when fitted) should be done just prior to the ignition checks rather than afterwards. This is because in the event of carburettor icing being experienced, a significantly higher mag drop can occur giving the pilot a misleading impression of magneto/engine malfunction.

Again on the subject of carburettor heat or alternate air systems, it should be borne in mind that either of these are alternate air systems. Thus, should a pilot find upon re-selecting cold air or the normal induction system that the rpm decreases, it will be indicative of a collapsed air filter or one which is partially choked by insects etc. – either of which will have a significant effect upon the take-off performance.

Finally, before leaving the subject of power checks it might be useful to note that in relation to battery charge and the ammeter indications the modern type of alternator will (provided the battery is in reasonable condition) top the battery up very quickly after starting. The result of this is that during the power checks little more than a zero indication may be shown by the ammeter. At this stage many pilots switch on the landing lights or similar system to get a reaction from the instrument and assess the charge state from this. However it is probably a better check to note the ammeter indication immediately after starting, when a significant and positive charge will be shown if the alternator is working correctly.

Pre-take-off checks

The subject of habits and their relationship to pilot competence and decision-making has already been mentioned earlier in this book. What perhaps needs to be restated is that habitual actions without thought can invite many unnecessary risks.

Perhaps it is in the aftermath of their initial training that many pilots fall into the habit of the 'numbers game', or the simple action of following the check-list by rote, which causes them to carry out actions without applying real thought to what they are doing. For example, during the preflight checks they synchronise the heading indicator with the magnetic compass without giving thought to whether the heading indicator has precessed markedly since it was synchronised with the compass in the parking area before taxying. Remember, the instrument checks during taxying only serve to indicate that the compass and heading or turn instruments are operating in the correct sense; these checks do not demonstrate whether an unusual amount of precession is taking place.

A good heading indicator will not normally precess more than approximately 10 degrees during the taxying period, and therefore a large variation between the compass heading and that of the gyro heading indicator just prior to synchronisation during the pre-take-off checks is a warning to the pilot that he may have a problem in that the instrument may give large errors during flight. Errors of this nature which have gone unnoticed have been responsible for pilots becoming unsure of their position, including inadvertent penetration of controlled airspace and danger areas, etc., thus exposing themselves and others to needless risks.

The use of flap during take-off

The use of a small amount of flap during take-off may, in certain circumstances, be advantageous, but most normal take-offs are conducted without flap being used. As a general rule the use of flap under normal conditions is a case of 'swings and roundabouts', in that whereas the use of flap will lower the stalling speed and hence the speed at which the aircraft can be safely lifted off the ground, the ensuing rate of climb will be impaired because of the increased drag.

In this respect it should be appreciated that different types of flaps produce advantages and disadvantages in varying degrees when used in the take-off situation. Therefore it is not possible to give a single hard and fast recommendation as to their use when it comes to the take-off phase of flight. Suffice to say that the manufacturers, quite apart from the theories and opinions often expressed by pilots, will undoubtedly have done their flight trials in a fairly sophisticated way, spending considerable time and money, with the object of demonstrating the best performance from their product. Thus if a single rule could be given it would be 'follow the manufacturers' recommendations as outlined in the aircraft manual'.

There are of course exceptions to the rule, though even these may be covered by information given in the aircraft manual. For example, the short field take-off which relates to clearing an obstacle ahead along the take-off path, and the soft field take-off which relates to the various conditions of soft ground, long grass or a rough surface.

The considerations applicable to the short field take-off have already been covered earlier in this book so the following comments only relate to the use of the soft field take-off procedure. It is during a take-off in this situation that the use of flap can without doubt be beneficial in that with flap lowered the aircraft can be safely lifted off the ground at a slightly lower airspeed.

Although in most cases this speed will be only some 5 knots lower it does have important benefits:

(1) The shocks on the landing gear when taking off increase their severity by the square of the speed. It will be most important to bear this in mind because the nose wheel could quite easily come into contact with the ground at a late stage of the take-off run when taking off from a rough surface. Thus even a small reduction in the lift-off speed will be an advantage from a structural point of view, particularly when taking off in conditions of little or zero headwind when the associated ground-speed will be higher than usual.

(2) An aircraft will not fly until it attains sufficient lift. This lift is achieved by an appropriate airspeed and angle of attack. Assuming the correct lift-off speed is 60 knots it can be seen that if the retarding influences of the surface, such as very soft ground or long grass, prevent the speed

from ever increasing beyond 55 knots, the aircraft will never become airborne.

However, it could be that with judicious flap selection the lift-off speed becomes 53 knots and now the aircraft will be able to become airborne. Nevertheless having made this statement it would still be wise to bear in mind that the aircraft manuals rarely contain more than an approximate indication of how soft ground or long grass will increase the length of the take-off run; after all, how soft is soft? – and how long is long? Therefore when using the soft field technique a pilot must ensure that the take-off run available is reasonably long and also be prepared to abort the take-off should things look as though a safe take-off in the available take-off length is doubtful.

The most dangerous situation is to try to get the aircraft off the ground below the lift-off speed, because although a desperate tug at the control column or a convenient bump on the surface may cause the aircraft to become airborne below the lift-off speed, this will usually only be a temporary condition in which safe and stable flight cannot be attained and the aircraft will stall or just drop back to the surface. In either case the pilot is very close to a disastrous situation.

The message is thus very clear – weigh up the advantages of using flap (if any) and lift off at the correct speed compatible with the aircraft's weight and the flap position selected. If there is any doubt that the correct lift-off speed will be achieved – **Abort the take-off**. What is fundamentally clear in these circumstances is that so often a sufficiently high risk element exists, making it a 'no-go' situation. Whether steps are now taken to reduce the weight of the aircraft by dispensing with the passengers, or whether the take-off is delayed until a significantly stronger headwind prevails, or whether both or other criteria must be met are decisions which any pilot may, from time to time, expect to be faced with. On such occasions the gambling instinct inherent in many of us must be suppressed.

A note of warning in relation to soft field take-offs is that it would be most unwise to attempt a take-off under conditions which give rise to the need for a soft field and a short field procedure to be combined. The reasons for this are fairly obvious in view of the foregoing comments.

Finally there is the question of whether it would be better to select the required flap position prior to take-off or wait until a late stage in the take-off run. Those pilots who adopt the latter procedure do so on the premise that with flap down the total drag on the aircraft will be increased and it would be better to suffer this extra drag for the shortest time possible during the take-off run. However, the amount of increased drag as a result of having a small amount of flap selected during the ground roll will be fairly small because most light aircraft have fairly low lift-off speeds.

If we consider this fact in a take-off situation which is going to be so tight that this small increase in drag is really significant then a 'no-go' situation already exists. Added to which one has to appreciate that any flap malfunction during selection, whether done manually or through an electrical system, will automatically make the situation even more critical. Thus with these thoughts in mind it becomes clear that the wisest course of action is to first ensure that there is a sufficient length available and then select the appropriate amount of flap before commencing the take-off run.

(Note: Comments concerning the procedure for raising flap after take-off or during a go-around situation are covered later under the 'approach and landing' considerations.)

Crosswind factors during take-off

A significant number of take-off accidents occur dur to the pilot losing control of the aircraft because of a marked crosswind. Additionally, during a crosswind take-off the headwind effect can be appreciably reduced leading to a pronounced increase in the length of the take-off run and distance.

Loss of control on the ground in crosswind conditions is mainly in the directional sense and usually caused by insufficient awareness and/or slow reactions or over-reactions by the pilot. Tail wheel aircraft are particularly vulnerable to crosswind effects whereas nose wheel aircraft are more directionally stable whilst on the ground. Nevertheless it would appear that nose wheel aircraft figure disproportionately in those accidents related to loss of directional control during both take-off and landing. This could be due to some pilots having a false sense of security when flying nose wheel aircraft, plus the fact that of recent years more and more aircraft movements are taking place at small landing strips which are often fairly narrow and lack the obstacle-free area of a normal aerodrome.

The type of accident which occurs under this heading, like so many others has three major contributing factors:

A lack of awareness.
A tendency to over-react.
Insufficient motor skills to cope with the situation.

In relation to all three it should be understood that the manufacturer, during flight tests, establishes a 'demonstrated crosswind component' up to which the aircraft is known to be safely controllable. This is in accordance with aircraft certification requirements laid down by aviation authorities. This demonstrated figure must be at least 0.2 Vso.

Although the demonstrated crosswind component which appears in the aircraft manual or on placards in the cockpit is not usually an actual limitation figure, any pilot who takes off or lands in crosswind conditions which give rise to a greater crosswind component than that demonstrated by the manufacturer is in effect acting as a test pilot in that he is operating the aircraft outside known and established parameters, and whilst it is true to

THE TAKE-OFF – PROCEDURES AND CONTROL

say that different pilots have different degrees of skill it is also true to say that when the wind is strong it is usually gusty in nature and sudden gusts can change the wind direction by 20 or 30 degrees and raise the wind speed by 40% or more.

To appreciate the effect of the vagaries of gusty winds during a take-off, or for that matter a landing, let us assume that the demonstrated crosswind component for the aircraft being used is 10 knots, a figure fairly common to many light aircraft. The reported wind velocity is 330/20 knots and the take-off runway is 30, giving a crosswind angle of 30 degrees. By reference to Fig. 5.4 we see that by entering the graph at the 30-degree angle point (A) and moving down to the 20-knot speed arc (B) the aircraft will be taking off in a crosswind component of 10 knots, which is just on the manufacturers' demonstrated figure, and the pilot should therefore be able to control the aircraft during the take-off run.

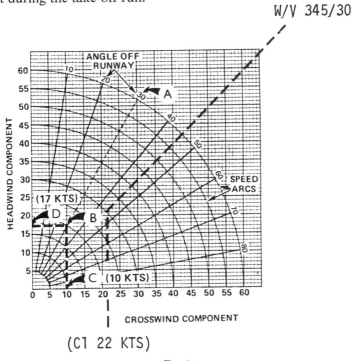

Fig. 5.4

However, if during the take-off a gust is encountered which changes the wind velocity to 345/30, a very different crosswind component will be experienced, as shown at (C1). From this it can be seen that a sudden gust which changes the direction and speed of the wind could easily change the situation into one in which the crosswind component becomes more than 100% greater than that in which the aircraft has been shown to be safely controllable. In these circumstances, and regardless of the skill of the pilot, the aircraft may quite well be incapable of being controlled and an accident is a reasonable certainty. Bearing these facts in mind it is clear that considerable caution must be exercised in weather conditions where strong wind gusts are being experienced.

So much for lack of awareness and skill; turning now to the issue of the pilot over-reacting, this error will tend to be greatest when using small and narrow landing strips. One has only to picture the situation during a take-off from a small strip with a hedge forming one boundary of the take-off path and a ploughed field or one in which wheat, several feet high, is growing, forming the other. No doubt there will also be a few trees at random intervals along the side bounded by the hedge.

It is in this situation that a pilot will be acutely conscious of the need to maintain accurate directional control over the aircraft and if a sudden gust occurs causing a change in the aircraft heading there will be a strong tendency for the pilot to use very positive rudder application to the extent where over-controlling can occur. Thus it is in these circumstances that any flaws in the pilot's motor skills will be magnified and, due to the critical nature of the take-off environment, an accident can quite easily occur.

Clearly the only sensible answer to the reduction of risk is to limit operations from sites of this nature to weather conditions more appropriate to the particular take-off environment.

Apart from these points it should also be appreciated that when a significantly strong crosswind is present during take-offs in nose wheel aircraft it will be advisable to keep the nose wheel on the ground a little longer than normal in order to obtain better directional control during the ground run. Additionally it may also be advisable, particularly during gusty crosswind conditions, to lift off at a slightly higher speed so that a more positive lift-off is achieved, one resulting in a smaller chance of the aircraft settling back onto the ground due to the reduction in speed if a gust should suddenly die.

Finally, a few brief extracts from recent accident summaries are reproduced below, and although pithy in nature they do illustrate how easily these occur when the pilot has not given enough thought to effects of a crosswind on take-off.

> 'During the take-off run in crosswind conditions the aircraft swung to the left and crossed the runway edge. The throttle was closed and as power reduced, the aircraft swung further to the left and then slid sideways damaging the landing gear, wing tip and propeller...'
>
> 'During a crosswind take-off the aircraft swung to the right and collided with a wire fence adjacent to the grass runway. The aircraft was extensively damaged...'
>
> 'The pilot over-corrected a swing on take-off and the left undercarriage collapsed. A strong gusty crosswind was present at the time...'
>
> 'During crosswind conditions the aircraft swung on take-off as the

aircraft was accelerating. As full rudder could not control the swing, take-off was abandoned. Although control was regained the tail rose and the aircraft nosed over ...'

'The aircraft was taking off from a narrow uneven strip between wheat fields. The pilot lost directional control and then groundlooped. Damage was consistent with the undercarriage and wing hitting the wheat crop ...'

'In conditions of a strong crosswind the aircraft struck a small bump during take-off. It became airborne in an excessively nose-high attitude as it passed the crest of the runway's uphill slope. The pilot attempted to regain control but the aircraft drifted to the right and landed heavily on the main wheels causing the support structure to collapse ...'

Wake turbulence

Turning from operations conducted at small aerodromes and landing strips to those which take place at larger aerodromes where a mix of small and large aircraft occur, all pilots must bear in mind the possibility of encountering the turbulent wake of large aircraft, particularly during the departure and approach and landing phases of the flight.

In earlier days it was thought that this turbulence was created by the propeller slipstream; however, tests have proved that slipstream effects dissipate to almost nothing within some 75 metres of its source, whereas the vortices created at high angles of attack produce very powerful rolling

Fig. 5.5

moments which will easily exceed the roll control capability of a light aircraft and last for a significant period. These comments must not be taken to mean that the effects of slipstream from propeller aircraft or helicopters can be ignored but the dangers associated with slipstream are usually more important when the aircraft is taxying or running up engines on the ground. This also applies to 'jet blast', the effects of which can be seen from Fig. 5.6. What is not always appreciated is the strength of these exhaust velocities in relation to the distance from the generating aircraft when 'idle power' or 'breakaway thrust' is being applied.

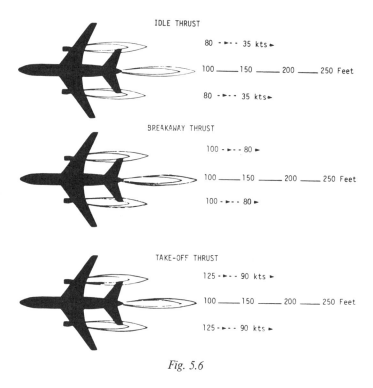

Fig. 5.6

The figures given in Fig. 5.6 clearly show the necessity for small aircraft to hold well clear of the active runway during pre-take-off procedures. Pilots must therefore be constantly aware of the hazards involved when taxying in the vicinity of large aircraft and helicopters, and give them a very wide berth.

Added caution is necessary when light aircraft approach or position at holding points prior to take-off and in such circumstances they must position well clear of larger aircraft, particularly when those aircraft are parked parallel to the taxyway.

The following paragraphs give a brief explanation of wake turbulence, how it occurs and how to avoid it.

THE TAKE-OFF – PROCEDURES AND CONTROL

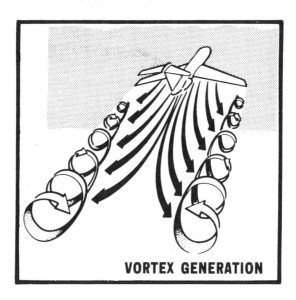

Fig. 5.7

All aircraft generate horizontal vortices from the wings whilst in flight and these can become very powerful at high angles of attack such as during lift-off or during the landing flare. These vortices are caused by the overspill of air from under the wing into the area of low pressure above the wing; the greater the pressure differential between the underside of the wing and the upper surface, the greater will be the vortex strength. Figure 5.8 shows the commencement of the vortex as air spills over the wing tip at (a). The airflow continues aft of the wing tip to produce two counter-rotating cylindrical vortices as shown at (b) and (c). The tangential velocity of the air within these vortices can exceed 150 kts and any small aircraft coming within them would be uncontrollable.

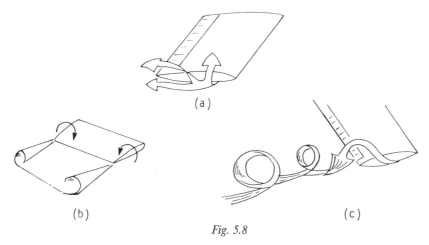

Fig. 5.8

The actual circulation of these two cylinders of air moves outwards and downwards behind the generating aircraft, and remains active for several minutes. With this knowledge, certain procedures can be used to avoid them.

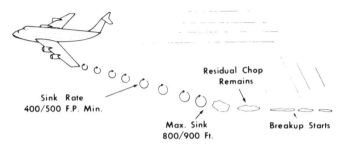

Fig. 5.9

Figure 5.9 shows the normal route taken by the vortices after being shed from the wing tips, and it can be seen that during flight it is safer to be slightly higher when passing behind or moving into the area previously occupied by a large aircraft.

Figure 5.10 shows the vortex movement near the ground in still air, viewed from behind the generating aircraft.

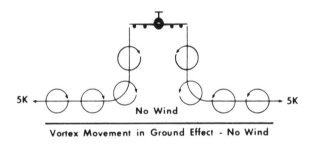

Fig. 5.10

The maximum tangential airspeed in the vortex system, which may be as much as 300 ft/sec immediately behind a large aircraft, decays slowly with time after the passage of the aircraft and eventually drops sharply as the vortex system disintegrates.

A crosswind will decrease the lateral movement of the upwind vortex and increase the movement of the downwind vortex as shown in Fig. 5.11. Thus a light wind could result in the upwind vortex remaining in the touchdown zone of a runway for several minutes longer than normal.

When taking off behind a large aircraft, delay take-off for at least two minutes and preferably longer. Rotate and lift off before the point of the departing aircraft's lift-off. Intersection take-offs would be inadvisable in these circumstances. Figure 5.12 shows a vertical and plan view of the vortex areas.

THE TAKE-OFF – PROCEDURES AND CONTROL

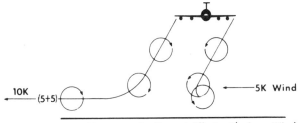

Fig. 5.11

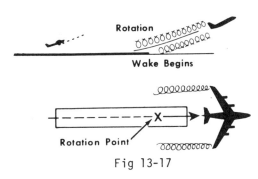

Fig. 5.12

Vortex wake generation begins when the nose wheel lifts off the runway on take-off and continues until the nose wheel touches down on landing.

If there is a crosswind, stay upwind of the large aircraft's climb path until turning clear of its wake. Avoid subsequent headings which will cross below and behind a large aircraft (Fig. 5.13). Be alert for any critical take-off situation which could lead to vortex encounter.

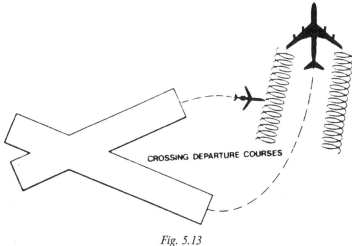

Fig. 5.13

Table 5.1 gives minimum times to allow before departing following the take-off of a larger aircraft.

Table 5.1 Wake vortex spacing mimima – departures

Leading aircraft	Following aircraft		Minimum spacing at the time aircraft are airborne
Heavy	Medium, small or light	Departing from the same position	2 minutes
Medium or small	Light	Departing from the same position	2 minutes
Heavy (full length take-off)	Medium, small or light	Departing from an intermediate part of the same runway	3 minutes
Medium or small (full length take-off)	Light	Departing from an intermediate part of the same runway	3 minutes

Wake turbulence – approach and landing

When conducting an approach and landing behind a large aircraft which has just departed or landed the following precautions should be observed:

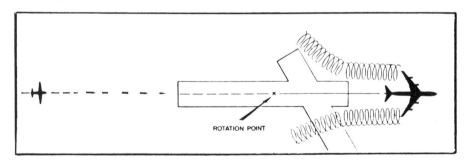

Fig. 5.14

- Landing behind a departing large aircraft. Note the large aircraft's rotation point – and if the large aircraft rotates when past the intersection you should ensure you land before this point. However, the time and distance separation as shown in Table 5.2 should be adhered to. If the large aircraft rotates earlier, as shown in Fig. 5.14, it would be advisable to discontinue the approach and remain above the departing aircraft's flight path. The minimum time and distance figures shown in Table 5.2 will still apply.

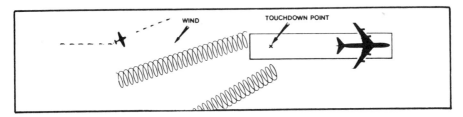

Fig. 5.15

- Landing behind a large aircraft which has landed ahead of you. Note the point of its touchdown and whether or not a crosswind is present, stay above the large aircraft's flight path and land beyond its point of touchdown (see Fig. 5.15). Observe the separation figures given in Table 5.2.

Table 5.2 Wake vortex spacing minima – final approach

Leading aircraft	Following aircraft	Separation minima distance and time equivalent			
		ICAO		UK	
		nm	min	nm	min
Heavy	Heavy	4	–	4	2
Heavy	Medium	5	2	5	3
Heavy	Small	–	–	6	3
Heavy	Light	6	3	8	4
Medium	Medium	3	–	3	2
Medium	Small	–	–	4	2
Medium	Light	4	–	6	3
Small	Light	–	–	4	2

These minima to be applied when an aircraft is operating directly behind another aircraft, and when crossing behind at the same altitude or less than 1,000 ft below.

The following occurrence report shows the importance of avoiding wake turbulence encounters. The aircraft concerned in this case was heavier and approaching to land at a higher airspeed than most light aircraft, yet a fatal accident was only just avoided.

Fig. 5.16

Severe wake turbulence encounter

Aircraft: Piper PA31 Navajo Chieftain
Date: September 1984

The pilot was on the approach behind a Boeing 737. At the outer marker he reduced speed to 110 knots to give a greater separation. The approach was on the centre-line and glideslope with only light atmospheric turbulence. At 300 ft he was cleared to land; landing flap was set and speed reduced to 95 knots crossing the threshold. Suddenly at about 25 ft a severe buffet was experienced and the aircraft rolled violently to the left through 25 degrees to 30° until application of full aileron, rudder and asymmetric power controlled the roll. The aircraft had been rolled and yawed clear of the runway. He climbed away from the ground, experiencing two more slight buffets with the rolling effect diminishing in strength. A normal landing was made further down the runway.

The pilot felt that had an overshoot been initiated when the buffet was

THE TAKE-OFF – PROCEDURES AND CONTROL

first experienced the aircraft would have climbed above the vortex. His company training highlights the problems of wake turbulence and suggests a high approach path to a non-limiting runway as one solution. In this case it was not possible as ATC had issued a 'land after' clearance with the B737 clearing slowly two-thirds down the runway.

The pilot considers that he was fortunate in that:

- the PA31 Chieftain ailerons are powerful,
- the aircraft was well below maximum landing weight,
- the piston engines have a quick response, and
- he used a high Vat of 95 knots (appropriate to 7,000 lb) on the long runway at Manchester airport.

If doubt exists on the separation necessary to avoid wake turbulence, a pilot should forego the approach. Whenever air traffic controllers are aware of the likelihood of a wake turbulence hazard, the phrase *wake turbulence* will normally be given, but pilots should note that because of the difficulty in predicting the occurrence of hazardous vortices this caution may not always be transmitted.

(Note: If the foregoing procedures conflict with the aircraft performance requirements, e.g. by limiting runway length, or with specified aircraft operating procedures, these will naturally take precedence and wake turbulence avoidance should be ensured by allowing additional time and distance separation.)

Birdstrikes

Birds, particularly seagulls at certain times of the year and day, find runways attractive places to rest and are viewed by many pilots more as an inconvenience than a hazard. Due to the relatively low lift-off speeds of light aircraft, pilots often assume that the birds will hear the noise of an aircraft engine, get airborne, and fly away from the aircraft's take-off path. What is often not appreciated is that aircraft normally take off into wind and when a stiff breeze is present the engine noise is not always clearly heard when upwind from the aircraft, and therefore on occasions birds will not become airborne until collision avoidance is impossible.

Birdstrikes, particularly with the larger varieties, can be extremely dangerous, as the following two summaries from accident reports indicate.

Summary no. 1: jet aircraft

'... A multiple birdstrike was experienced at 50 ft. No. 1 engine surged and the throttle lever slammed rearwards by itself passed the detent and unlocked the thrust reverser. The engine was shut down, and a single-engined landing made.

No. 1 engine nose cowl was missing, eight first stage fan blades were liberated and the inlet case and both front and rear fan containment cases had major penetrations. Two of the engine mount bolts were fractured and the engine was only attached by the front left cone bolt and flexible lines at the rear of the engine. Bird remains were found in the fan discharge duct, LH main gear well and outboard trailing edge flaps of the RH wing. Three bird carcasses were found on the runway.

Birds were black-headed gulls which weigh 275 g; *one* bird entered the engine, striking no. 20 fan blade 2″ from its root; it broke off, initiating the engine failure. Debris exited at the 8 o'clock position and broke the outer pane of the flight deck side window and entered no. 2 engine, causing blade and stator damage.'

Summary no. 2: piston engine aircraft

'... The aircraft, a light twin-engined Dove, became airborne and then collided with a flock of seagulls. At the time of the collision the landing gear retraction had just been initiated and later several birds were found trapped in the landing gear bays. The aircraft sustained damage to the ailerons and these became practically immovable, with the aircraft in a left wing down attitude. The rudder was used to manoeuvre the aircraft and during this stage the left engine lost power and within a minute it stopped completely.

Directional control was, by this time, marginal, but the pilot was skilful enough to land the aircraft successfully within $2\frac{1}{2}$ minutes of the birdstrike occurring. Examination of the aircraft showed fuel leaking from the left tanks caused by a bird having struck and opened the water drain cock beneath the wing. Another gull was found jammed between the underside of the left wing and the aileron mass balance forcing the aileron into the up position.

Both engines had all air intakes clogged with bodies and feathers. The de-icing boots on the propellers and wings were damaged and the carburettor air filter of the right engine was smashed. The airframe had suffered damage in many places including the flaps. One herring gull had penetrated the leading edge of the right wing as far back as the main spar and a total of 14 identifiable gull bodies were removed from the aircraft.

Although in this case it was not possible to know of the existence of the birds before take-off because the flight started when it was still rather dark just about dawn, the moral we are left with is very clear. In general, and even in normal daylight conditions, birds can be rather slow in taking any avoiding action so pilots should take care not to get themselves into situations in which a high risk of birdstrikes exists.'

The following analysis of birdstrikes to general aviation aircraft below 5,700 kg has been made by J Thorpe of the UK CAA Safety Data and Analysis unit and certainly shows the need to be constantly aware of such collisons during the take-off and initial climb, and the approach and landing phase of any flight.

Analysis of bird strikes to UK general aviation aircraft, 10-year period (Aircraft of weight less than 5,700 kg)

(1) In the ten years covered by the study there were 400 reported strikes. This is about one-seventieth of the rate for aircraft of over 5,700 kg.
(2) The birds struck are similar to those struck by the large aircraft group (in brackets): gulls 54% (53.2), lapwings 18% (12.1), pigeons 9% (6.0), swallows 3% (3.6). Two large birds, a greylag goose 3.2 kg and a gannet 3.5 kg, were struck, and in each case very considerable structural damage was done.
(3) 50% of strikes to light aircraft occurred during landing, and 44% during take-off and climb. These percentages are identical to those for the large aircraft group. During cruise, 4% of strikes occurred, compared with 2.1% on large aircraft, almost certainly due to the lower cruise altitudes used by light aircraft.
(4) The aircraft speed at which the birdstrikes occurred differed significantly from large aircraft: 55.3% of strikes were below 80 kts, compared with 20.6% on large aircraft. It was noticeable that 21% of strikes were below 60 kts – it may be that the quieter and smaller general aviation aircraft do not provide the same alert response in birds.
(5) 85% of strikes were between ground level and 200 ft, with only 1% above 2,500 ft. These figures are similar to the large aircraft group.

Comparison of parts struck by birds, general aviation and large aircraft

	General aviation aircraft %	Large aircraft %
Fuselage	6	13
Nose/radome	11	32
Windscreen	11	14
Engine/propeller	23	19
Wing	33	16.5
Landing gear	11	5
Empennage	3	1

(6) The parts struck were very different from large aircraft, the wing accounting for 33% of strikes, compared with 16.5% in large aircraft.
(7) On large aircraft only 18% of strikes cause damage, whereas on general aviation aircraft 23.5% of strikes cause damage. The major effects were skin denting in 12% of strikes. However, 9 incidents involved structural deformation. There were 4 cases of windscreens being smashed, 2 of

them causing minor injuries to the pilots. A gyrocopter was ditched in the sea when severe vibration was experienced immediately after passing through a flock of gulls.
(8) The monthly variation is almost identical to that on large aircraft. August is the worst month for strikes, the same as for large aircraft, the flying peak coinciding with the bird population peak.

Flocks of birds parked on or adjacent to a runway or take-off area are not always easy to cope with. The carriage of a 'bird scare gun' in the cockpit is hardly practical, sensible or even legal, and taxying up a runway to scare birds away prior to take-off can often merely result in the birds returning to the same position as the aircraft is returned to the take-off point. However, there is one elementary precaution which can be taken. When birds are on or closely adjacent to the take-off path, be prepared to close the throttle immediately it becomes evident that the birds are being slow to move away to a safe distance from the aircraft's line of departure.

Reducing the risk during take-off

- Consider the condition of the propeller more carefully than perhaps you have done in the past. This does not mean placing your body or limbs within the propeller arc; it can be inspected, quite satisfactorily and with safety, by standing in front of it and behind it whilst running your fingers along the front and rear surface of the blades – remember, it is the tip sections which are most prone to damage.
- Use the correct check-list and conform to it, but nevertheless think about what you are using it for and consider your actions relative to the prevailing circumstances.
- If the limiting 'mag drop' is exceeded, either maximum or differential, you have a 'no-go' situation. When checking the idling rpm, remember that a faster than normal 'tickover' will increase the float period during the landing flare. This could lead to an accident if the wind is light and the landing area is short.
- During the take-off roll, carry out a check on the oil temperature, (cylinder head temperature if applicable), the oil pressure and the rpm achieved. This is the time to abort the take-off if any indications give cause for concern. If the engine fails during the initial stage of the climb-out you will have a major emergency on your hands and one which might have been avoided by monitoring the appropriate instruments during the take-off roll.
- Know your lift-off speed relative to the approximate weight of your aircraft. Abiding by this speed will significantly improve your take-off distance.
- Ensure the crosswind component is within the maximum recommended for the aircraft, and remember when crosswinds are present the headwind

THE TAKE-OFF – PROCEDURES AND CONTROL

component will be reduced and cause your gradient of climb to be reduced. Be alert for wind gusts, remember that in crosswind conditions an aeroplane can be likened to an airport baggage trolley with a will of its own as far as direction is concerned.
- When taking off from a small landing strip or similar area you should bear in mind that if you have passed the half-way point and the aircraft has not reached its lift-off speed you are now entering the unknown and taking a gamble with your life and the lives of your passengers.
- When operating from larger aerodromes the possibility of wake turbulence effects will often exist. On such occasions delay take-off in accordance with the figures shown in Table 5.1. Once caught in the horizontal whirlwinds generated by large aircraft it is unlikely that you will be able to maintain lateral control of your aircraft, even with full aileron deflection.
- During take-offs and 'go-arounds', treat birds which are on or near the flight path with extreme respect. Over 40% of birdstrikes involve seagulls and similar-sized birds. In-flight collisions with birds of this size can be catastrophic.

Quote from an AOPA (USA) Bulletin ...

'... in 1906 0.3 persons were killed for every 10 million cart miles travelled.
In 1980 0.32 persons were killed for every 10 million aeroplane miles. Thereby hangs a tale ... General aviation and old dobbin are about on par when it comes to safety in transportation!'.

6
The approach and landing – procedures and control

The three primary causes of accidents and incidents during an approach and landing appear to be as follows:

(1) Attempting to land when the distance available is inadequate for the aircraft in the conditions prevailing at the time.
(2) Failure to position the aircraft at the right place, at the right time, at the right height and with the right speed for the approach, even though the landing distance is adequate.
(3) Inability to cope with the existing crosswind conditions, or turbulence.

The problems arising from (1) above have been covered under the heading of performance accidents, thus leaving (2) and (3) to be covered in this section. Of these, the comments under (2) appear most frequently in the accident records and are primarily related to lack of judgement, skill, or care in planning the approach. If we refer back to Table 2.1 it can be seen that during the 5-year period shown in this table a total of 487 accidents occurred in the approach and landing phase of flight – which clearly shows that the largest number of accidents occur during this period. What can also be noted from the accident reports is that, apart from the need to understand the 'performance factors' which affect the aircraft, it is during this period that a pilot's motor skills and his ability to make quick and correct decisions can be heavily taxed.

In considering how safety can be improved during this phase it is necessary to appreciate that in general terms the question of reducing the risks often starts when the aircraft is being initially prepared for the arrival phase at the aerodrome, e.g. during the aerodrome approach checks. Most pilots carry out these checks fairly well but often forget to prepare themselves mentally for the entry to the traffic pattern and the conduct of the final approach. For example, giving serious thought to the amount of traffic which might be in the circuit, the types of aircraft using the aerodrome, the weather conditions including the degree of crosswind (if any), listening out on the 'tower' or airfield information frequency (as and when applicable), and similar considerations. It is through these actions that a safe pilot prepares himself by being forewarned of the various activities and conditions which he might have to cope with as he enters the traffic pattern.

The aerodrome approach checks which relate to physical actions on the part of the pilot are in essence a method whereby the pilot establishes certain facts about the aircraft such as fuel state, accuracy of the heading indicator, frequency selection (when radio is used), altimeter setting, etc. Thus when carrying out the 'prelanding checks' at a later stage, most of these will be of a confirmatory nature (excluding such items as lowering the landing gear), thus enabling the pilot to give more attention to such matters as lookout (it should be borne in mind that approximately 85% of midair collisions occur within traffic patterns), controlling the aircraft's position in the circuit and weighing up the conditions to be expected during the final approach.

It is in relation to the execution of these functions and the preparedness of the pilot's thought processes that it might be useful to consider again the pilot's attitude of mind and the development of habits. After all, habits start from the time we are taken out of our nappies and sat on a pot and a continued development of habit training occurs from then on. If we now relate this to a pilot's early flying training we should appreciate that pilots are weaned in a fairly sheltered environment – that of a training aerodrome, where it is normal to have an adequate and safe landing distance on almost all occasions. However, this will not necessarily apply when using landing strips and some small grass aerodromes, and thus pilots trained in rather forgiving surroundings can easily run into serious problems later in their flying activities unless they reach and maintain a high standard of proficiency in accuracy landings.

Now, whether that old adage 'practice makes perfect' holds up in aviation parlance may be open to debate, but one statement which must be true is, 'a pilot will not improve his performance without physical practice'. The moral in this case being for pilots to have a complete understanding of the problems which might occur during an approach and landing and then to habit-train themselves to carry out accuracy landings whenever possible regardless of the available length of the landing area. Being 'too hot and too high' on an approach is no excuse for endangering the safety of an aircraft or its occupants.

Aerodrome approach checks

Accidents rarely 'just happen'; they usually begin to happen as a result of events which lead up to them. It is for this reason that so many accidents need never have occurred if the situation and the chain of events had been recognised by the pilot at an earlier stage. How does this explanation fit in with the aerodrome approach checks? Well, just take the following three examples.

- Fuel tank selection
  A fairly common move by pilots when approaching the destination aerodrome is to first put the aircraft into a descent where little and sometimes no power is being used. During this descent the aerodrome approach checks are often initiated, in the course of which the fuel

selector may be changed to a different tank. If this action is carried out improperly, e.g. mis-selection to the OFF position, selection to the tank with a minimal fuel content, or not fully turned to the correct position, it is unlikely that the pilot will find this out until he increases the power for level flight at the bottom of the descent. At this stage the aircraft will most likely be at or around 1,000 feet above the ground and the pilot will actually have created his own engine failure situation in circumstances which will leave very little time for a remedy to be effected, particularly if the propellor stopped during levelling off from the descent (and make no mistake, this has happened to a number of pilots). Such a situation would not occur if the pilot ensured that any fuel tank selection was made during the normal cruising flight and the descent initiated a minute or so afterwards, because any mis-selection of the fuel cock when the engine is under cruising or higher power will be brought to his notice within a few seconds of the mistake being made.

- Altimeter settings

  Another common action is to re-set the altimeter datum during the descent prior to entering the traffic pattern. Bearing in mind the importance of having the correct altimeter setting during poor weather conditions it is clearly preferable that any change in setting should be made prior to, rather than during, the descent phase. Actions such as altimeter adjustments, fuel re-selection, etc., should be double-checked after a brief interval to ensure they have been carried out correctly and because the workload can be significantly increased during the descent towards the destination airfield. It is advisable to avoid the need for increasing it further at this time, and therefore the aerodrome approach checks should be completed, and double-checked where necessary, *prior* to initiating the descent.

- Heading indicator – synchronisation

  For similar reasons the resynchronisation of the heading indicator with the magnetic compass should be included in the aerodrome approach checks and completed prior to descending. Additionally, as the aircraft gets nearer to the surface it will, on many occasions, become more difficult to hold it steady enough for an accurate compass reading to be taken – apart from the fact that it will normally be slightly less reliable when the aircraft is in the descending attitude. Past incidents have shown that pilots have misinterpreted the active runway during their joining procedure and this has led to conflicts with other aircraft in the circuit; some of these occasions were caused by the pilot having set the heading indicator incorrectly.

From these examples we can see that pilots have the ability to guard against mistakes but they have to use this ability at the right time in order to avoid situations conducive to accidents. This comment also applies to the prelanding checks which are, of course, conducted prior to any landing in the interests of safety. Bearing in mind that the circuit area is one in which a high risk of collision between other aircraft exists, and the fact that lookout is what is left of 'look-in' time, the prelanding checks need to be restricted to essential actions and should be performed without the use of a check-list. Allowing for the need to keep the prelanding checks as short as possible compatible with their purpose, it has been noted that not all pilots include a check of the brake pressure (on those aircraft equipped with hydraulic brake systems) or the engine temperature and pressure. Including these two items may, one day, bring very significant benefits – for example, noting the fact that there is little or no resistance on the pedals alerts the pilot to an equipment unserviceability which could affect his stopping distance (this applies particularly to a small landing strip) or his ability to control the direction of the aircraft during the roll out following a crosswind landing. A pilot who is alerted in this manner is in a good position to make a prudent decision as to whether to continue with the planned landing, proceed to a larger landing field, or, in the event of a strong crosswind, use a different runway. Again, in the event of a red line indication on the oil pressure gauge the pilot will be alerted to the fact that in these circumstances a glide approach landing should be carried out rather than a powered approach.

Lookout in the circuit

Fig. 6.1 (Photograph courtesy of UK Accident Investigation Branch)

FLIGHT SAFETY IN GENERAL AVIATION

As stated earlier, some 85% of midair collisions occur in the circuit pattern. Therefore, during this phase of flight 'lookout' becomes even more vital. The need to maintain a vigilant search for other aircraft will clearly exist throughout the whole circuit period but there are a couple of very important considerations which all pilots must bear in mind if they are to reduce the risk of a collision with another aircraft.

During the downwind leg a pilot's attention is occupied with a number of items and high up on the list of those requiring attention is the need to complete adequate prelanding checks whilst maintaining the aircraft's track parallel with the runway in use. A pilot cannot look out and 'look in' at the same time, but during the prelanding checks a certain amount of 'look-in' time is required. So check-lists should not be used at this stage and the list of checks should not be allowed to become too long, e.g. such items as synchronising the heading indicator with the magnetic compass and resetting the altimeter datum should already have been done during the pre-take-off checks if carrying out a series of circuits, or during the aerodrome approach checks if a cross-country or local area flight has been conducted.

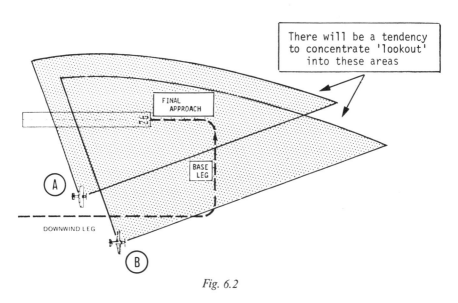

Fig. 6.2

With regard to the maintenance of the aircraft track relative to the runway in use, it is very easy to pay so much attention to this that lookout in the opposite direction to the runway becomes neglected. Figure 6.2 illustrates the dangers of this. Consider for the moment the dangerous situation which can be created if the pilot in aircraft 'A' fails to maintain a lookout to the right side of the aircraft. If the pilot in aircraft 'B' fails to see the other aircraft because it is outside his visual search scan then there will be a high risk of a near miss or midair collision, particularly if aircraft 'B' turns onto the base leg before aircraft 'A'. If we now add to this the possibility of a height disparity between the two aircraft the whole situation is worsened, and even more so if a mix between a high wing and a low wing aircraft is involved (see Fig. 6.3), and bear in mind that these blind spots will always be there, right up to the point of touchdown.

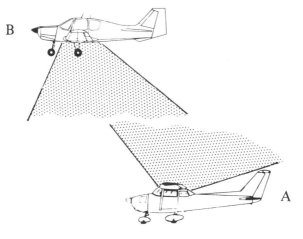

Fig. 6.3

Fig. 6.4 Two students and two instructors luckily survived this collision which took place when their blind spots coincided at an uncontrolled airport. The Cherokee, making a steep close-in approach, descended on top of the Cessna, already on final.

Setting up the base leg

An old and true axiom is that a good landing stems from a good approach. This means having the aircraft at the right height, airspeed, configuration and position relative to the runway, and in the case of visual circuits this starts on the base leg.

Examination of past landing accidents reveals that in numerous cases the pilot was 'too hot and too high' – the aircraft was 'in command' and the pilot failed to make the right decision to go round again, or made this decision too late. (This latter error being the most fatal mistake when using small aerodromes, and particularly landing strips.)

Being too hot and too high is usually the result of a badly planned or executed base leg in which the aircraft is too close to the landing path at the commencement of the base leg turn, or is carrying too much power during the initial stages of the base leg. It is not always easy to determine how the pilot gets into this situation but nevertheless there is a strong link between this type of accident and the rather automated power control procedure which many pilots are taught in the initial stages of their training – a habit which many pilots retain even after gaining the experience which permits proper judgement, rather than the 'numbers game', to be used. The expression 'numbers game' refers to the habit of initially using a fixed amount of power for every base leg on every occasion. To understand why this is not good practice one has only to consider the effects of aircraft position, height and wind during the base leg phase.

If a pilot is sent further downwind by an air traffic controller, or has to go further downwind because of an aircraft ahead, simply setting the fixed power setting (say 1,600 rpm) used when turning at the normal position, will only result in a long dragged out low approach, particularly if the headwind is strong. In which event the effects of turbulence will be increased and prolonged, to say nothing of creating an unnecessary noise nuisance to people below the aircraft's flight path.

If, on the other hand, a pilot is asked by air traffic control, or for some other reason is forced, to turn onto base leg earlier than usual, then the use of the fixed rpm will create a tendency for the aircraft to become too high, and a 'hot and high' situation is likely to result. To those who might argue that the pilot can always alter the power setting if the aircraft becomes too high or too low one must point out that hot and high accidents normally occur because the pilot takes this action far too late or is continuously slow in making corrections. Thus it is better to be alert to the varying amounts of power which may have to be selected at the commencement of the base leg rather than use a fixed power setting, which often proves to be substantially wrong.

So much for the situation of an early or delayed base leg turn; what should be considered next is the effect of wind. Consider for a moment the pilot who has a tailwind component along the base leg. Selecting the usual

THE APPROACH AND LANDING – PROCEDURES AND CONTROL

fixed rpm will automatically result in the aircraft becoming too high, and thus the tendency for a high approach will occur at a time when the headwind component on final approach will be rather less because of the crosswind effect. This, once again, could easily lead to a 'hot and high' situation as the pilot approaches short finals!

An automated power setting, coupled with a headwind component along base leg will call for a positive increase in power just prior to the turn into final approach. This could lead to a low airspeed during the turn – remember, a significant amount of flap will normally have been selected by this stage. In any event an unnecessarily low approach, with all its disadvantages, is quite likely to occur. Clearly, a better procedure will be to plan and select a power setting compatible with the aircraft's position and height relative to the landing path, and include an allowance for the prevailing wind. From which point on, the pilot should be alert to the gradient of descent and make early adjustments as necessary. This is the most sensible and effective method of initiating an accurate approach path.

The precision approach

Landing accidents generally fall into several categories, i.e., overruns, undershoots and loss of directional control, or control in pitch, after touchdown. The first two are usually accompanied by a compulsive determination on the part of the pilot to continue with the landing even though all his senses should be telling him to go round again. This also often applies to the third category because in many instances the real cause of loss of directional control begins before touchdown as a result of being too hot and/or too high, or attempting to land in unsuitable conditions.

This information is relatively simple to obtain by merely reading through the accident reports, and reducing the risk of being involved in landing accidents is not particularly difficult either. It simply requires the pilot to give more thought to the planning of every landing and be critical of the way he implements it. This will lead to a greater sense of awareness, an improvement of judgement and an increased ability to control the aircraft safely.

Therefore, whenever approaching the landing aerodrome or strip, take into consideration the important factors such as, the difficulties likely to be experienced, e.g., the landing site location and the degree of mechanical turbulence which might be experienced on the approach. Landing sites which are in close proximity to trees or undulating terrain can produce a fair degree of turbulence or that 'sinking feeling' on the approach, even though the air at cruising altitude is quite smooth.

Think about the length of the landing path relative to the aircraft being used and the strength of the headwind component on final approach. During the downwind leg assess the likelihood of any crosswind effects, during both the approach and the touchdown. If the decision is made to increase the threshold speed because of anticipated turbulence remember that this can increase the float period significantly, particularly if a crosswind condition is prevailing. Failure to appreciate how an extra 5 or 10 knots at the threshold creates a long float period under crosswind conditions has led to many a disastrous arrival.

During the final approach do not relax even when things seem to be going well; monitor airspeed and power – early corrections should be your aim as these mean smaller corrections which add to the preciseness of the approach.

Much has been written over the years as to whether, in propeller-driven aircraft, the elevators control the airspeed whilst power controls the rate of descent or whether power controls the airspeed and elevators control the rate of descent. However, as the relationship between power and elevators during a descent is the same as that which exists between the ailerons and rudder during all normal flight manoeuvres, the situation is one in which whatever you do with one will affect the other, so it is surely more a matter of thinking philosophy rather than separate control applications. Therefore you would probably be best advised to use the method you have been taught – but to use it properly.

If you doubt the foregoing comments and believe that power controls the airspeed, then just as a touch of light relief it is suggested that you climb up to 2,000 feet, set cruising power, trim out, and when the airspeed has stabilised, reduce the power whilst leaving the elevators alone; the nose will lower and the airspeed will increase. Now return to normal level flight and then increase the power without moving the elevators; the nose will rise and the airspeed will decrease. You have now proved you are correct – but not perhaps in the way you meant!

To return to the final approach and more serious matters, bear in mind that a sensible pilot will, above all, make the decision to go round again at a sufficiently early stage to ensure clearance from obstructions at the far end of the landing path or in the initial climbout area. This often means making an earlier decision when using a small landing strip. This comment does not mean that an early decision may not also be necessary at some of the smaller aerodromes used by general aviation aircraft.

In this respect, a useful reminder to all pilots is that when the decision to go around is made late and on the basis of there being insufficient runway remaining in which to stop the aircraft, it will be extremely doubtful whether go-around action can be undertaken if there are obstructions close to the end of the landing path – thus in these circumstances it may often be wiser to continue with the landing run and accept the smaller risks involved in a low speed collision with obstacles rather than the high risk situation of stalling at a low height in attempting an impossible climb over obstructions, or having an airborne collision with them.

Whatever philosophies a pilot may hold in relation to go round again procedures, the need to make this decision early will arise from time to time

THE APPROACH AND LANDING – PROCEDURES AND CONTROL

Being wise afterwards
---------- is too late!

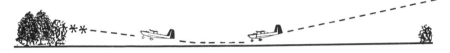

Fig. 6.5

and it is on these occasions that a late decision or delayed action may produce anything from a hair-raising experience to a total disaster, as Fig. 6.6 mutely shows.

Fig. 6.6

Use of flap

It is difficult to postulate fixed rules with regard to the use of flap during an approach and landing, because of the differences which exist between types of aircraft, types of flap and the varying conditions which may be encountered. However, running over a few facts relating to the effect of flap when lowered will clearly be of value to the eventual decisions concerning when, where, and how much flap to lower in particular circumstances.

Firstly, the reason why flaps were introduced and fitted to aircraft was due to advances made in relation to aerodynamic design – advances which resulted in streamlined aircraft with very low drag characteristics.

Due to the high drag values of early aircraft pilots had to adopt a very low nose attitude in order to maintain a safe speed whenever a significant reduction in power was made. This became an advantage during an approach and landing because it gave the pilot a good view ahead along the descent path.

In modern aircraft the low drag inherent in their design resulted in very flat glide angles and therefore during an approach it was difficult to provide the pilot with an unobstructed view of the landing path. The introduction of flaps overcame this problem by creating more drag and thus the aircraft nose could be lowered to a steeper angle without increasing the airspeed, giving the pilot a better view along the approach path without having to suffer the disadvantage of excess airspeed at a time when this would be undesirable.

Over the years different types of flaps evolved, from the simple to those which increased the wing area and/or created slots, thus increasing the amount of lift for a given airspeed and angle of attack. These improvements also lowered the stalling speed, enabling an aircraft to fly more slowly without degrading safety.

Today there are many types of flaps and combinations of flaps in use and the amount by which they can be lowered also varies between aircraft. As for their benefits, apart from providing a better view ahead, the decrease in stalling speed when flaps are lowered is clearly a valuable one. However, the actual reduction in stalling speed from the clean condition to that in which the flaps are fully lowered is relatively small and this fact is not always appreciated. For example, using the figures given in the manuals of twelve different current light aircraft reveals that the reduction in stalling speed when flaps are fully lowered varies between 3 and 9 knots, and when taking an average of these figures it comes to 6.5 knots. Therefore the impression that the use of full flap will appreciably reduce the length of the landing roll is not necessarily true. If one uses the average figure obtained from this survey, purely as an academic example, it shows that the difference between landing with no flap and with full flap is equal to the difference between landing in a zero headwind and one of 6.5 knots. This does not imply that full flap should not be used for normal landings, it merely means that, in relation to the landing roll, full flap gives the greatest advantages during zero or light winds and when the wind is strong this advantage decreases.

This simple fact might affect your decision to use full flap when landing in very strong wind conditions. Why? Well, just consider in what situation the use of full flap may be a disadvantage... First consider the possible effect of having a lot of drag when approaching to land with a very strong wind straight down the runway. In such conditions, and particularly when the surface below the approach end of the runway slopes down, the wind gradient and shear effects could cause strong downdraughts to be present. These could be of the order of 100 to 250 feet per minute, or more.

Now, consider the aircraft: most light aircraft have about 50% excess thrust horsepower available for climb performance, i.e., light aircraft are usually able to maintain altitude down to 50% of their horsepower; any power setting below this will normally cause the aircraft to lose altitude. The 50% excess thrust horsepower available usually results in a rate of climb between 500 to 700 feet per minute. However, if the flaps are lowered fully, a large amount of this excess horsepower will be needed to overcome the drag produced by the flaps. All pilots know how the climb performance is degraded during a go-around with full flaps lowered.

Now, assuming the aircraft you are flying has 50% ETHP which gives a rate of climb of 500 fpm, then this roughly equates to 100 fpm for every 10 ETHP available. If full flap produces a drag value which takes 40 ETHP to overcome (often the case in aircraft which have a 40-degree or more flap deflection), this will leave only 10 ETHP for the climb performance, in which case the maximum rate of climb in the full flap condition will be approximately 100 fpm. Therefore if the pilot runs into conditions on the approach in which a downdraught exceeds this figure (a far from uncommon occurrence) then the only direction the aircraft will go, even with full power applied, is down – a fact that a number of pilots have discovered when their undershoot accident occurred!

Additionally, it is in these circumstances that there will be a strong tendency for the pilot to raise the aircraft's nose and increase the angle of attack to the extent that it is flying on the backside of the power curve, as explained later on page 97.

Turning to the situation where a strong crosswind component exists during the approach and landing, there is one factor concerning the operation of flaps which at times tends to get overlooked, yet might have an important bearing on the amount of flap a pilot decides to use during the approach.

This is the fact that the use of flap often leads to a small decrease in control effectiveness of the ailerons, elevators and rudder. One result of lowering flap is to decrease an aircraft's lateral stability and aileron effectiveness; another is that the lowered flap deflects the air flow and slipstream effects further downward and below the tailplane. This leads to a reduction in the effectiveness of the elevators and rudder for a given airspeed and power setting.

In considering these facts, together with the lower stalling speed achieved by flap deflection, the reduction on control effectiveness could become a major consideration in the amount of flap to use for a landing when a strong crosswind component is present. In other words, bearing in mind the need to have maximum control effectiveness available to combat turbulence and align the aircraft along the runway centreline just prior to touchdown, it can be seen that a surefeit of flap might be the causal factor for an accident to occur in these conditions.

Fig. 6.7

Having made this point, there is no hard and fast rule in relation to procedures or handling techniques to be used when strong crosswinds are present. It all depends upon the situation at the time, the length and width of the landing path, the actual landing site, e.g. the degree of turbulence to be expected, the type of aircraft, and finally the experience and ability of the pilot.

Nevertheless, what can be stated is that flaps have been provided for the pilot's use, but the decision as to when, and how much, can only be left to the pilot and he must make this decision according to the circumstances at the time. In this respect just bear in mind that the manufacturer has also provided you with a throttle, a mixture control, an alternative or hot air source for the engine, and most probably a cigar lighter. All these and many other items are made available for the pilot's use, but having them available does not mean that the throttle must be fully open or fully closed on the approach, nor does it mean that the mixture control should be in the 'lean' position at all times. (Come to that, nor does it mean that you must use the cigar lighter when you are a non-smoker.) So it would be of value for pilots to relate this philosophy to their decisions and use what they require in accordance with their knowledge and common sense in relation to each and every flight situation.

Having considered some aspects of lowering flaps there are also some points to stress concerning the raising of them, points which might help pilots in reducing the risk factor.

Following a take-off and particularly a go-around with flap down, most pilots use an arbitrary height at which to raise the flap. This habit, formed during training, becomes a relatively automatic procedure. However, what needs to be understood is that the airspeed at the time of raising flap can be vitally important.

Some pilots are under the impression that when raising flap during the initial stage of the 'climb out' the aircraft will automatically sink or experience a loss in the rate of climb – and this is quite true if the aircraft is handled improperly. What is equally true is that provided a sufficient airspeed is achieved before raising flap then the angle of attack can safely be increased as the flap is raised and no sink or reduction to the climb rate will occur.

Whilst this may seem a little academic bearing in mind that the drag will be reduced when flap is raised and thus the airspeed increased for the same aircraft attitude, the important factor is that the raising of flap in these circumstances tends to become an automatic action and if undertaken without reference to the ASI, the airspeed could be low enough to cause a marked sink to occur, with the result that the pilot might instinctively move the control column backwards thus lowering the airspeed still further and bringing the aircraft close to the stall.

In any event, if this happens the aircraft will be in a situation where it is flying close to or within the region known as the 'backside of the power curve' which is also referred to as the 'region of reversed command'. To understand this it is necessary to refer back to basic principles of flight and appreciate that any speed below the endurance speed will be one where the aircraft is flown at an angle of attack in which the drag becomes greater and more power will be needed to overcome it or the aircraft will sink. During the initial 'climb out' the power will normally be at a maximum and the only way to rectify the situation will be to lower the aircraft nose and increase the airspeed, an action which will of course result in a height loss.

Figure 6.8 illustrates the effect of flying an aircraft in the speed region below endurance speed. The endurance speed is shown at 'A' and this is the point at which the least power will be required to maintain the aircraft in level flight. Any attempt to maintain height at a lower airspeed will require a large increase in power or the aircraft will lose height. Should the aircraft be flown so that it is operating on the backside of the power curve during the initial climb from a take-off or a go-around situation, the airspeed could decay very rapidly if the pilot moves the control column back to avoid any sink as flap is raised, and thus a stall is quite likely to occur – and at a height from which recovery cannot be effected.

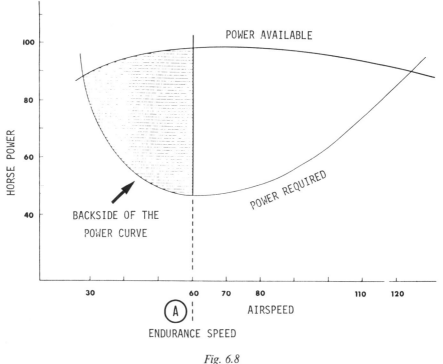

Fig. 6.8

The danger of flight in this region is naturally of equal importance during the approach and landing phase. Care must be exercised to avoid this particular condition when winds are strong, severe turbulence exists and full flap is lowered for the approach.

The following pithy extracts from recent accident reports highlight the use of wrong procedures and the failure of pilots to control the aircraft properly during the landing phase:

> 'During a practice forced landing into a field, an overshoot was initiated at about 30 ft. The aircraft failed to clear the trees. The pilots escaped with cuts and bruises, but the aircraft was destroyed. It was found that the flaps were fully deployed ...'
>
> 'There was a high tailwind from the left and following the initial touchdown the aircraft veered to the right into a ground loop. The undercarriage legs folded up as the bungee support system failed ...'
>
> 'Approaching to land with full flaps, the aircraft developed a very high sink rate. Although full power was applied the aircraft struck the ground heavily. The pilot attempted to overshoot but the aircraft failed to gain height and struck two parallel hedgerows ...'

THE APPROACH AND LANDING – PROCEDURES AND CONTROL

'During the landing run on a short runway the pilot realised that the aircraft would not stop, so he initiated an overshoot. In an attempt to lift the aircraft over obstructions it stalled and crashed in the field beyond ...'

'The pilot landed deep into the runway length and floated some way before touchdown. With the runway end approaching fast he decided to go around. Full power was applied and the aircraft was lifted off before a ditch at the end of the runway but hit an earth mound which wiped off the undercarriage ...'

'Following touchdown half-way down the runway after approaching at higher than normal airspeed due to turbulence, the aircraft struck a fence during a go-around, sustaining damage to the tailplane and landing gear door hinge ...'

'The co-pilot was briefed to touch down some 475 yd from the end of a grass strip. After the initial flare the aircraft continued to float, the pilot in command considered an overrun into trees would result, so took control and initiated an overshoot. The engine did not develop maximum power immediately and as the aircraft climbed away it struck the surrounding trees at 10–12 ft below the tops ...'

'The pilot landed the aircraft well into the runway over high trees. The aircraft bounced and floated, after which the pilot decided to carry out a go-around. The aircraft failed to get airborne in rough ground beyond the runway and swung to the left. The throttle was closed and the aircraft hit a ditch, ripping off the main landing gear ...'

'In conditions of light wind and after a prolonged float, the aircraft touched down half-way along the runway. When it came apparent that a safe landing could not be completed the pilot overshot but the aircraft struck the boundary hedge and the left wing detached ...'

'After landing half-way along the strip the pilot decided to overshoot. There was little response from the engine and the aircraft went through a barbed wire fence and struck a stone wall ...'

'On landing, the aircraft floated more than was anticipated. The pilot attempted to overshoot but the aircraft struck the boundary fence ...'

'The pilot had visited the strip by road to examine it from the ground before flying in with another pilot as passenger to fly out another aircraft. The pilot flew low over the airstrip and then made another approach. It touched down well inside the boundary, moving fast and skipping across the surface. At a late stage, power was applied and the aircraft became airborne in a steep nose-high attitude. The aircraft passed through the tops of some trees and then sank, hitting power lines before crashing into a sewage works ...'

'A swing developed during a crosswind landing, during the overshoot a wheel struck a dry stone wall on the airfield boundary causing the aircraft to pitch down into a small field. To avoid a second wall, the pilot groundlooped the aircraft ...'

'Control was lost during a crosswind landing and the aircraft came to rest in a ploughed field south of the runway. The nose gear, engine and the wing tip were damaged ...'

'During a crosswind landing the aircraft swung off the runway on landing and hit an airport fence ...'

'The pilot in a non-radio aircraft failed to notice the runway change. At the end of the landing run in a crosswind, the aircraft groundlooped to the left, damaging the landing gear ...

'The aircraft landed in a crosswind and nosed over, causing damage to the undercarriage and propeller ...'

Fig. 6.9 (Photograph courtesy of Bedford County Press)

Sooner or later all aircraft stop flying; when this happens you had better be in the right place at the right time and in full control of the situation.

Reducing the risk during the approach and landing

- Whenever possible, and provided a long en-route descent to an aerodrome is not being implemented, always change fuel tanks when in level cruising flight.

THE APPROACH AND LANDING – PROCEDURES AND CONTROL

- Avoid the common error of staring at the magnetic compass when resynchronising the gyro heading indicator. There is always a strong tendency to watch the compass as an initial action instead of maintaining the wings level and the heading steady with the aircraft in balance. In the past pilots have approached or landed on the wrong runway due to using incorrect heading indicator re-setting procedures.
- Unless necessary, a pilot should avoid changing fuel tanks during the prelanding checks. If a mistake is made it is highly likely that it will not be discovered until too late.
- Do include a 'brake pressure' check during the aerodrome approach or prelanding checks, otherwise you may discover that one or both brakes have become unserviceable just when their need is vital.
- Pay particular attention to the area on the outside of your circuit; remember the runway will not get up and hit you, but the aircraft which you have not seen most certainly can.
- Before turning onto base leg, carefully check the area inside and behind your circuit path. This will mean using your body, your neck and your eyes.
- On base leg avoid selecting power 'by numbers'; choose a power setting which takes into account the effect of the wind and the distance from your chosen touchdown area. It is usually on the base leg that the 'too hot and high' situation starts to develop.
- Plan each and every landing you make – and then pay constant attention to making the approach as precise as possible. Be critical of your own performance throughout the entire approach.
- A good rule when increasing the approach speed to allow for gustiness and turbulence is to add one half of the gust speed to the appropriate approach speed, i.e. if the wind is gusting from 15 to 25 knots the gust speed is 10 knots, therefore add half this to your normal approach speed. However, the maximum advisable increases to the approach speech for a light aircraft is 10 knots. This is because a higher figure will significantly extend your float period, leading to control problems when winds are strong or turbulence creates difficulties.
- During approaches at small aerodromes and landing strips, increased speeds can lead to unacceptable float periods and increase the risk of an accident, an important point to bear in mind.
- In many light single-engine aircraft the stalling speed variation between having the flaps down or up and flying at light or gross weight is relatively small – therefore do not be over-zealous in adding to your approach speed if you are contemplating a landing at all-up weight permitted and a lesser amount of flap than usual. Study your aircraft manual and obtain the necessary information so that you can make the correct addition to your approach speeds.
- When significant obstructions exist along the climb out area of the runway being used, your decision to go around again must be made early.

It may not be possible in the existing conditions to go around after a bad landing has been made, especially if touchdown is made well up the available landing path. It is in situations like this that a precise approach must be made if you are to remain within the safety envelope.

- Study the weather conditions in relation to your landing area and your aircraft and use flaps in a sensible fashion – that flapless or full flap landing you were exhorted to use at a different aerodrome, in a different aircraft, under different conditions, a year ago, may not be the best one for this occasion.
- If a sudden and positive sink occurs during the approach it may be necessary to lower the aircraft's nose to increase your airspeed (and thus your lift) as well as making a positive increase to your power setting. If your speed is already low, it is possible to get into the situation where you are flying on the backside of the power curve at a time when height is at a premium and recovery to safe flight impossible.
- Many landing accidents occur because the pilot relaxed and stopped flying the aeroplane in the final stages of the landing. Therefore stay alert throughout the flare and ground roll phase.

7
Fuel management

Engine failure or malfunction accounts for some 25% of general aviation accidents, and high up on the list of factors which cause an engine to stop is fuel mismanagement. Once airborne a general aviation pilot is busy enough controlling the aircraft, scanning the flight instruments, monitoring the gauges, maintaining a lookout, responding to Air Traffic Control instructions, trying to remain within his privileges, and abiding by the Rules of the Air and other regulations. Add to this the workload involved in navigating the aircraft, looking after the passengers, and thinking ahead, and it can be seen that on many occasions a pilot has quite enough to do without volunteering for other tasks.

Nevertheless, in view of the number of times in the past that the engine has stopped and led to an accident, the extra task of fuel management is more vital than many pilots appreciate. The correct way to manage your fuel situation is to start on the ground during preflight planning. Take the trouble to read and understand the information given in the aircraft manual, both in relation to fuel consumption figures and how the fuel system should be operated. Armed with this knowledge the question of fuel management, once airborne, simply becomes one of monitoring the fuel consumption at various stages of the flight, coupled with the use of correct fuel system operating procedures.

In considering the activities involved in correct fuel management we could divide general aviation flight operations into two types:

(1) Operations which are conducted within the local flying area, during which the aircraft is either in the circuit or fairly close to the base aerodrome.
(2) Navigation flights in which the aircraft operates at some distance from an aerodrome, and one in which the vagaries of the weather can cause many well-planned flights to go awry.

Complacency

Past accident reports have shown that in the situation described (1) above and particularly when conducting circuit operations, many pilots have demonstrated a complacency with regard to fuel management. This complacency is largely brought about by the fact that if fuel runs low during circuit operations the pilot can simply terminate the flight earlier than intended – no sweat. In other words the pilot places his life and those of his passengers in the hands of the fuel gauges which, sadly, have often let down such a trust. Add to this the fact that during circuit flying the aircraft spends approximately half its time in the air below 500 feet, a height band which, in the event of an engine failure, is the most dangerous place to be in. So although fuel management is vital during cross-country flights it is also equally vital to apply it properly to any flying operation and the comments which follow regarding cross-country flying and fuel management basically apply to any type of flight.

When planning a cross-country flight it is not sufficient to just fill the tanks and go. It is essential that the pilot knows not only the actual fuel state before departure but also the anticipated fuel consumption en route. There is also a further aspect which is not always considered and that is how to operate the aircraft and its fuel system to achieve the anticipated consumption figures. Sadly this is not always given sufficient emphasis during basic instruction when the various training constraints will interfere with the practical application of mixture control. So often during a pilot's initial training the question of fuel management merely boils down to filling the tanks prior to departing on a cross-country flight, and this leads to the development of bad habits both in preflight and in-flight procedures.

Misunderstanding

Information regarding fuel consumption figures is given in the aircraft manual but there are at least two commonly misunderstood facts about the figures which are quoted and these are:

(1) The figures shown in the aircraft manual are based not only upon using appropriate power settings but also on the use of the mixture control system, so pilots should bear in mind that the use of the mixture control lever is not just confined to placing it in 'rich' to start or 'idle cut-off' to close the engine down.
(2) The use of carburettor heat to combat or prevent carburettor icing can cause an increase of up to 15% in fuel consumption for a specific power setting.

If a light aircraft is operated during cruising flight with the mixture control in the full rich position it can result in a reduction of at least 1 hour or more from the figures quoted in the aircraft manual. This could mean a reduction in range of from 100 to 150 nm and if during the flight it becomes necessary to use carburettor heat for a significant period the range and endurance figures will be further reduced.

The accident statistics use two terms when referring to fuel-related causes of engine failure. These terms are:

FUEL MANAGEMENT

Fuel exhaustion – As it implies, this means that all the fuel had been used up.

Fuel starvation – This means that there was fuel available but for some reason it was not reaching the engine.

Lack of care

Fuel exhaustion appears far too frequently as a cause of accidents, and with the exception of those cases which occurred due to the pilot becoming lost, the main reason for this is carelessness, lack of knowledge, or simply lack of proper preflight planning. Fuel starvation on the other hand is primarily due to lack of knowledge of the fuel system, incorrect cockpit procedures before or during flight, or basic mistakes by the pilot in his practical handling of the fuel system, and before discussing this aspect further it would be appropriate to remind the reader of the following:

> There is no pilot born, who has not reached out at some time or other and got hold of the wrong knob, switch or lever, or, having got hold of the right one, turned or moved it in the wrong direction, or has simply forgotten to reach out at all. This applies to us all, whether we have a few hours or thousands of hours in our log books!

Just knowing this fact and keeping it in mind when we fly will help us to remember to double-check our actions on appropriate occasions and these occasions mean from the time prior to take-off to the time of arrival at the destination aerodrome. It should also help us to plan our operating procedures in a fail-safe manner whenever possible.

Think fail-safe

How does this last statement relate to the take-off phase and the application of good fuel management? Well, consider for a moment the pilot who selects a fresh fuel tank after his power checks have been completed and immediately prior to lining up for take-off. In this case there might not be a cloud in the sky and the visibility could be 50 kilometres, yet the pilot will be taking off into the unknown and could be at a very high risk indeed. Why? Because the fuel line from the newly selected tank may be restricted or blocked. Again, the pilot may have just inadvertently selected the wrong tank, e.g. an auxiliary tank which can only be used safely when the aircraft is in level flight, or even placed the fuel selector lever into the OFF posiiton. In any of these events the engine will stop at the most dangerous time – during a late stage in the take-off run, or just after becoming airborne.

You may rightly say, why did the engine not stop before or during the period the throttle was opened for take-off? A good question, and the answer to it is simply that even if the selector was moved into the OFF position there will still be fuel left in the line from the tank which had been used previously, and thus the engine will not stop immediately but continue running for a period which depends upon the rate of flow demanded from the engine and the length of the fuel line. The actual time involved with many light aircraft can be up to two minutes at idling power and may be just over ten seconds at full power. In either case there may be sufficient time for the aircraft to become airborne before the engine fails and/or insufficient runway remaining in which to stop the aircraft before it reaches the aerodrome boundary.

How can a fail-safe procedure be planned to guard against this event? Simply by carrying out the engine run-up whilst using the tank which will be used for the take-off. Therefore, in those aircraft fitted with alternate tank selection systems the full procedure would be to start up and taxy on one tank and then select the other just prior to conducting the engine run-up. In this way both tanks will have been proven to be feeding before flight.

However, bear in mind that the first tank selected will not have been checked to ensure if it is feeding properly at more than low power, and therefore once the aircraft has reached a safe altitude after take-off it would be advisable, particularly when conducting a cross-country flight, to re-select the initial tank and continue on that tank to ensure that it is feeding properly at cruising power before getting too far from the departure aerodrome, thus ensuring that all the usable fuel carried will be available to you during the flight.

Another aspect of fuel management during the take-off phase is the correct use of auxiliary fuel pumps (when fitted). Most high wing aircraft equipped with low power engines merely rely on a simple gravity feed fuel system. If there is fuel in the tank(s) and the lines are undamaged or unrestricted the only thing that can prevent fuel flowing to the engine is the disappearance of gravity, in which case the pilot would become capable of sustained flight without the use of power, as indeed would the rest of the earth's population.

Nevertheless, certain high wing aircraft with higher performance engines, and all low wing aircraft, are fitted with auxiliary fuel pumps. In low wing aircraft an engine-driven mechanical fuel pump is necessary to raise the fuel to the carburettor and an auxiliary electric fuel pump is also fitted to cater for the possibility of the mechanical pump failing. For safety reasons it is a sensible precaution to switch on the auxiliary pump during those periods of flight when a sudden loss of power would be critical, i.e. take-off or approach to land. For this reason the auxiliary fuel pump should be switched on before take-off and left on until a safe altitude has been reached, at least 1,000 feet above ground level. Some aircraft are fitted with an auxiliary fuel pump which has two positions, a 'high' and a 'low', in which case the aircraft manual will include instructions in its use

If a fuel pressure or fuel flow gauge is fitted this should be monitored for a short while after switching the pump off to ensure that the fuel pressure is being maintained. If the fuel pressure or flow decreases after the pump has been switched off it could be symptomatic of an engine pump failure and the auxiliary pump should be switched on again. The reason for leaving the pump on until reaching at least 1,000 feet is that in the event of the mechanical pump failing, the fuel lines will empty prior to the engine failing and even if the auxiliary pump is switched on immediately the engine fails, it will take time before the fuel lines are re-primed and the engine can restart.

Good fuel management is a measure of pilot performance but the word performance can mean many things and in the case of an aircraft is variously used to describe rates of climb, reliability, maximum airspeed, maximum altitude, low speed handling, etc., etc. In relation to fuel consumed it is usually interpreted as meaning 'range' or alternatively, 'endurance'.

Range and endurance

A pilot knows that the range and endurance figures for a given aircraft will be found in the aircraft manual, but have you considered how a pilot who has carefully worked out his fuel consumption figures for a given trip can easily upset his calculations when in flight? For example, having selected a good economical cruising airspeed he runs late on departure and in order to make up time he then decides to increase the cruising power. To see the results of this action let us look at some actual consumption figures applicable to a typical light aeroplane.

Reference to Fig. 7.1 shows that the aircraft, when flown in ISA conditions and with lean mixture, uses 4.3 gallons per hour at 2,000 feet and 53% power. If the flight was planned for this power setting and altitude the aircraft in question with full tanks would have an endurance of 5.6 hours and a range of 476 nm in still air, based upon a usable fuel quantity of 24.5 gallons. However, if the pilot now decides to open up to 75% power the endurance would drop down to 4 hours and the range in still air will become 400 nm. A further point is that at 75% power it is unlikely that lean mixture would be used, and therefore the approximate endurance figure would actually be reduced to just over 3 hours and the range would be nearer to 300 nm. Thus a most significant reduction to flight safety would have occurred.

Taking another example from a different light aeroplane (Fig. 7.2), and using 4,500 feet as the cruising altitude, the endurance under ISA conditions and lean mixture at 56% power can be interpolated as 4.4 hours. However, if 75% power is used in the same conditions the endurance figure is reduced to 3.3 hours and if rich mixture is used the endurance could drop to 2.75 hours, a long way short of the 4.4 hours calculated originally.

Fuel quantity – the physical check

Another important aspect is that whilst it is clear that no amount of fuel management will help you if you have insufficient fuel left in your tanks, what is not so obvious to some pilots is the need to make a physical check

CONDITIONS:
1670 Pounds
Recommended Lean Mixture (See Section 4, Cruise)
NOTE:
Cruise speeds are shown for an airplane equipped with speed fairings which increase the speeds by approximately two knots.

PRESSURE ALTITUDE FT	RPM	20°C BELOW STANDARD TEMP			STANDARD TEMPERATURE			20°C ABOVE STANDARD TEMP		
		% BHP	KTAS	GPH	% BHP	KTAS	GPH	% BHP	KTAS	GPH
2000	2400	---	---	---	75	100	6.1	70	99	5.7
	2300	71	96	5.7	66	95	5.4	63	94	5.1
	2200	62	91	5.1	59	90	4.8	56	89	4.6
	2100	55	86	4.5	●53	85	4.3●	51	84	4.2
	2000	49	80	4.1	47	79	3.9	46	78	3.8
4000	2450	---	---	---	75	102	6.1	70	101	5.7
	2400	76	101	6.1	71	100	5.7	67	99	5.4
	2300	67	95	5.4	63	94	5.1	60	93	4.9
	2200	60	90	4.8	56	89	4.6	54	88	4.4
	2100	53	85	4.4	51	84	4.2	49	83	4.0
	2000	48	80	3.9	46	78	3.8	45	77	3.7
6000	2500	---	---	---	75	104	6.1	71	103	5.7
	2400	72	100	5.8	67	99	5.4	64	98	5.2
	2300	64	95	5.2	60	94	4.9	57	93	4.7
	2200	57	89	4.6	54	88	4.4	52	87	4.3
	2100	51	84	4.2	49	83	4.0	48	82	3.9
	2000	46	79	3.8	45	78	3.7	44	76	3.6
8000	2550	---	---	---	75	106	6.1	71	105	5.7
	2500	76	104	6.2	71	103	5.8	67	102	5.4
	2400	68	99	5.5	64	98	5.2	61	97	4.9
	2300	61	94	5.0	58	93	4.7	55	92	4.5
	2200	55	89	4.5	52	87	4.3	51	86	4.2
	2100	49	83	4.1	48	82	3.9	46	81	3.8
10,000	2500	72	103	5.8	68	102	5.5	64	101	5.2
	2400	65	98	5.3	61	97	5.0	58	96	4.8
	2300	58	93	4.7	56	92	4.5	53	91	4.4
	2200	53	88	4.3	51	86	4.2	49	85	4.0
	2100	48	82	4.0	46	81	3.9	45	79	3.8
12,000	2450	65	100	5.3	62	99	5.0	59	98	4.8
	2400	62	97	5.0	59	96	4.8	56	95	4.6
	2300	56	92	4.6	54	91	4.4	52	90	4.3
	2200	51	87	4.2	49	85	4.1	48	84	4.0
	2100	47	81	3.9	45	80	3.8	44	78	3.7

Fig. 7.1

of the fuel quantity before flight, even if it does mean climbing a stepladder for those high wing types. The reason for this physical check is that fuel gauges can go wrong and even when working normally they can have inherent errors which cannot be programmed out without additional expense in manufacture.

One has only to read the following extract from a General Aviation

FUEL MANAGEMENT

CRUISE & RANGE PERFORMANCE

GROSS WEIGHT 1560 LBS.
STANDARD CONDITIONS
ZERO WIND
LEAN MIXTURE

ALTITUDE	RPM	PERCENT POWER	TRUE AIR SPEED	GALLONS/ HOUR	ENDURANCE HOURS	RANGE MILES
2500	2600	86	136	7.4	2.8	379
	2500	78	130	6.6	3.1	404
	2400	71	123	5.9	3.6	433
	2300	64	116	5.3	3.9	449
	2200	58	108	4.8	4.3	460
	2100	52	99	4.5	4.6	456
4500	2600	82	135	7.0	3.0	395
	2500	75	129	6.3	3.3	418
	2400	67	121	5.6	3.7	441
	2300	61	113	5.1	4.0	453
	2200	● 56	106	4.7	● 4.4	458
	2100	51	96	4.4	4.6	444
6500	2600	79	134	6.7	3.1	407
	2500	72	127	5.9	3.5	432
	2400	65	119	5.4	3.8	446
	2300	59	112	4.9	4.2	460
	2200	54	104	4.5	4.5	464
8500	2600	75	133	6.3	3.3	426
	2500	68	125	5.7	3.6	440
	2400	62	117	5.2	3.9	454
	2300	57	109	4.7	4.3	459
10,500	2600	72	130	5.9	3.5	435
	2500	66	122	5.4	3.8	447
	2400	60	114	5.0	4.1	455

NOTES:
1. Range and endurance data include allowance for take-off and climb.
2. Fuel consumption is for level flight with mixture <u>leaned.</u> See Section III for proper leaning technique. Continuous operations at powers above 75% should be with full rich mixture.
3. Speed performance is with wheel fairings. Subtract 2 MPH for speed performance without wheel fairings.
4. For temperatures other than standard, add or subtract 1% power for each 10° F. below or above standard temperature respectively.
∗ 5. Cruise propeller is standard on TR-2. For TR-2's equipped with optional climb propeller use Trainer data and add 2 MPH.

Fig. 7.2

Safety Information digest to appreciate the need to visually check the quantity of fuel in the aircraft tanks...

  '... Both fuel quantity indicators showed full when the tanks were virtually empty, the cause being that the indicator assembly was loose in the pedestal mounting so that the earth wire from the electrically operated instruments was making intermittent contact, sending both indicators off scale to the full position ...'

Need one say more!

A lack of understanding of the fuel management problems involved in aircraft operation has been a fairly common cause of accidents throughout the years, as the following typical extract from a safety leaflet will show:

'... The aircraft had departed with fuel believed to be sufficient for the 1 hour 15 minutes flight. This belief was based upon the Flight Manual consumption figures at 2,000 feet and 2,200 rpm and an estimate of the fuel remaining before take-off from the elapsed time flown since the tanks were filled earlier in the day.

The flight took 15 minutes longer than planned because of ATC zone routings and during the approach the engine stopped, although the fuel gauges indicated 22–30 minutes' fuel remaining. The aircraft forced landed in a field but fortunately without damage. Upon investigating, the aircraft was found to be out of fuel.'

In view of the previous comments with regard to fuel consumption it is clearly dangerous to estimate fuel remaining by calculations based upon time flown by previous pilots since the tanks were filled. Small errors in the time calculations made for previous flights can add up to a significant error after several flights have been made, and it is often impossible to determine whether the tanks were actually full following a refuelling which had occurred earlier in the day.

In general, it could be said that the endurance of a light aeroplane is limited more by the endurance of the human bladder than the quantity of fuel in the aircraft's tanks. If so, it should be remembered that our bladders give us a physical warning that they are reaching their limit and this physical discomfort is far more accurate and 'attention-getting' than many aircraft fuel gauges.

Fuel indicating systems and inherent errors

Apart from the more obvious problem which can occur when a fuel gauge is over-reading, there are others. For example, the errors which may occur due to the construction of the fuel indication system. The fuel gauges used in light aircraft are normally of the direct reading type or are operated electrically. The direct visual reading type where the pilot can see the actual fuel level by looking directly through a sight glass are usually fairly accurate, though allowances do have to be made in the case of tail wheel aircraft due to the change of altitude when on the ground as distinct from the attitude taken up by the aircraft during cruising, climbing and descending flight. These same allowances also have to be made whenever fuel indicators are used in conjunction with a float-type mechanism in the fuel tank.

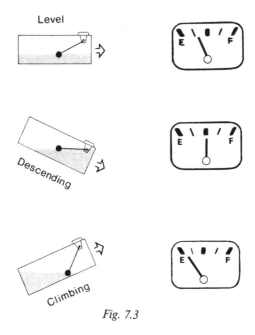

Fig. 7.3

However, many of the fuel indicating systems which are operated electrically use float-type mechanisms and these will only give accurate readings when the system is working properly and the aircraft is in level flight. Thus inaccurate indications of fuel quantity will be given when the aircraft is climbing or descending or turning, as shown in Fig. 7.3.

When the tanks are full or nearly full these errors will normally be relatively small but in the case of the tanks being less than one-third full the errors assume larger proportions. Whether the fuel gauges will over-read or under-read in a climb or descent will depend upon where the float mechanism is fitted in the tank, and therefore some aircraft will have gauges which over-read in the climb while others will under-read.

The importance of establishing the over-reading/under-reading characteristics of fuel guages on a particular type and model of aircraft can be better appreciated by considering the psychological effect on a pilot who takes off with the tanks one-third full. If he glances at the gauges just after take-off he may see indications that the tanks are empty. During a descent in the same aircraft with the same fuel state he could easily be misled into thinking his fuel state is greater than it is. Either of these two situations could lead a pilot into making an accident-inducing decision during flight.

Fuel tank selection

Another factor which has caused quite unnecessary accidents is the use of incorrect procedures when re-selecting fuel tanks during flight. Many of the fuel exhaustion accidents occurred because the pilot was not aware of the proper procedure, or failed to use it. Due to the differences between the correct procedures to use in the different types and models of general aviation aircraft it is not possible to cover this subject to any great extent in this book, and thus the main advice offered is to read and understand thoroughly the operation of the fuel system in the particular type of aircraft you intend to fly; there is no short cut to obtaining this kind of knowledge.

Furthermore, it should be appreciated that modifications to aircraft fuel systems, e.g. the fitting of long range tanks, auxiliary tanks, etc., are not uncommon, and this may not be revealed by simply reading through the manual issued with the aircraft when it was new. The possiblity of any changes which have been made should therefore be checked before flight.

A secondary item of advice in relation to fuel tank selection relates to habits formed when flying a specific type of aircraft for some time. Such habits may need changing when converting onto a different type of aircraft because of the variation in the selector positions between different types. In this respect and because of the natural development of habitual actions it must be understood that when a pilot is flying an aircraft in which he has had little experience it will be quite easy (particularly during times of increased workload) to move the fuel selector to the same position which he has been used to when flying a previous aircraft. This type of error can be aggravated if a tank which is low in fuel has been mistakenly selected. The error will not immediately be apparent because, even in the worst case, i.e. selecting the

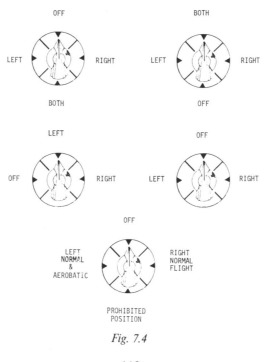

Fig. 7.4

FUEL MANAGEMENT

OFF position, the engine will continue running for a short while until the fuel in the lines has been used up. Therefore a mis-selection just prior to take-off or during the prelanding checks could have fatal consequences.

Therefore, to reduce the risk of selecting the wrong fuel selector position because of previously formed physical actions, you should positively double-check your actions whenever re-selecting a fuel tank. One has only to study Fig. 7.4 which shows some typical variations of fuel selector layouts as used in light aircraft to appreciate how a pilot could, without conscious effort, select a fuel valve to the wrong position. No pilot is immune from making this mistake and the double-check method is the best safeguard procedure.

Having covered some of the problems involved in fuel *quantity* the next step is to consider the fuel *quality*. In recent years a number of aircraft have been inadvertently refuelled with the incorrect grade of fuel and even Avtur instead of Avgas. It is the pilot's responsibility to ensure the correct grade and type of fuel is pumped into the aircraft tanks, and by monitoring fuel uplifts he will most certainly reduce flying risks.

Fuel contamination

Another far too common occurrence is fuel contamination, normally caused by the presence of water. Therefore pilots must diligently follow the recommended fuel straining technique given in the particular aircraft Flight Manual. This action assumes greater importance when an aircraft has been parked out overnight, in which case a larger amount of fuel should be strained.

The mechanism, operation and position of fuel strainers and drains will vary between different makes and models of aircraft and it will be necessary for a pilot to become conversant with these components of the fuel system for any aircraft he intends to fly.

The fuel strainer (Fig. 7.5) is situated at the lowest point of the system and is normally fitted either at the bottom of the firewall between the engine

Fig. 7.5

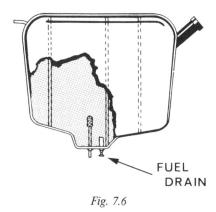

Fig. 7.6

bay and the cabin or sited underneath the selector valves with access from underneath the aircraft.

In some aircraft the fuel strainer is operated from within the cockpit and in this case it would be advisable to operate the strainer before commencing the external preflight inspection. This will enable the pilot to make a visual check to ensure that the strainer has been shut off properly. There have been cases in the past where aircraft have run out of fuel due to a strainer valve malfunction.

Other fuel drains are normally located at the bottom of each tank (Fig. 7.6) and they may be of the 'finger operated' quick-release type shown in Fig. 7.7.

Another method requires the use of a special tool with a probe which is pushed up into the drain release valve. In either case a water contamination

Fig. 7.7

check can only be made by collecting a fuel sample in a perspex or glass container. If any water is present it will separate from the fuel (Avgas 100LL is coloured blue) and sink to the bottom of the container. There is little point in operating a fuel strainer so that the fuel merely pours out onto the ground; this action will only prove the strainer system is working, it will not prove whether or not the fuel is contaminated.

To fully understand the importance of checking for any water contam-

FUEL MANAGEMENT

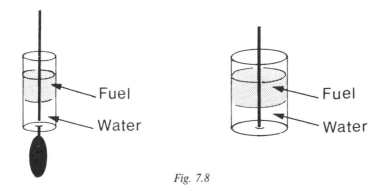

Fig. 7.8

ination in aircraft fuel systems one has only to read this extract from an FAA Safety Bulletin.

'... Three gallons of water were added to the half-full fuel tank of a popular make, highwing monoplane. After a few minutes, the fuel strainer (gascolator) was checked for water. *It was necessary to drain ten liquid ounces of fuel before any water appeared.* This is considerably more than most pilots drain when checking for water.

In a second test with the same aircraft in flying attitude (to simulate a tricycle geared model) the fuel system was cleared of all water; then one gallon of water was added to the half-full tank. *It was necessary to drain more than a quart of fuel before any water appeared.*

Fig. 7.9 Once is not enough – continue until there is no trace of water.

In both of these tests, about nine ounces of water remained in the fuel tank after the belly drain and the fuel strainer (gascolator) had ceased to show any trace of water. *This residual water could be removed only by draining the tank sumps.*'

Two significant findings emerged from the above tests and from tests made on a plastic mockup of a similar fuel system:

(1) When water was introduced into the fuel tank it immediately settled to the bottom, but did not flow down the fuel lines to the fuel strainer until all fuel was drained from the lines. Remember, each fuel tank must be turned ON to drain the tank lines through the gascolator.
(2) Since it was found impossible to drain all water from the tank through the fuel lines, it was therefore necessary to drain the fuel tank sumps in order to remove all water from the system.

Fuel tank venting

Before leaving the subject of fuel management it would be pertinent to introduce a few comments on the condition of fuel vents. All fuel tanks need to be fitted with some form of venting system in order that the ambient atmospheric pressure can be maintained within the tank as the fuel level lowers. If a venting system becomes blocked a suction will be created which will eventually overcome the action of a fuel pump (or gravity) and the fuel will not flow to the carburettor.

The method of venting varies with different aeroplanes; sometimes it is merely a hole drilled into the filler cap, other methods incorporate a valve within the filler cap or utilise a length of piping which terminates at some distance from the tank. Care must therefore be taken to ascertain the exact method used and it is vitally important to ensure that these openings to atmosphere are unblocked and undamaged, during the preflight inspection.

It should also be understood that a blocked vent will not necessarily reveal itself in the form of an engine stoppage when the tanks are fairly full. It takes time for the pressure within the tank to be reduced as a result of the suction generated by the fuel pump. In some cases a blocked vent has caused engine failure within a few moments of full power being applied and in

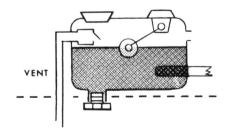

Fig. 7.10

others the fault has stopped an engine after a long period of flight, as evidenced by the following extract from an accident report:

'... At about 1,000 feet on the approach following a one-hour flight the port engine stopped, with all the symptoms of fuel starvation. The pilot landed safely on one engine and during the subsequent examination it was found that the port flexible fuel tank had collapsed due to a blocked fuel vent. This resulted in a large fold of the flexible material used in the tank construction blocking the fuel outlet pipe to the carburettor. It also created a fuel quantity indication of $\frac{3}{8}$ths full when in reality the tank was only $\frac{1}{4}$ full.

Frequency of powerplant failures

Tables 7.1 and 7.2 show the frequency of powerplant failures due to mismanagement of the fuel system, fuel exhaustion or other reasons. Allowing that some of the undetermined reasons for engine failure were probably associated with improper operation of the fuel system and some of the fuel exhaustion causes were due to inadequate fuel planning it can be seen that a significant number of engine failures were related to fuel, or rather the lack of it reaching the engine. Although Table 7.1 concerns only one year it can be noted that adding up the percentage of accidents due to 'powerplant failure for undetermined reasons' together with 'pilot mismanagement of fuel' and 'fuel exhaustion', we have the figure of 20.3% which is fairly close to the 24.9% figure for 'engine failure or malfunction' over the five-year period 1976–1980 as shown in Table 7.2.

The high frequency of accidents involving fuel and the powerplant indicates an area where pilots are exposing themselves to high risk.

Table 7.1 Most prevalent detailed accident causes, all fixed wing aircraft, 1981 (Annual review of accident data, US general aviation 1981)

Detailed cause	Number of accidents	Percent of accidents
Pilot–inadequate preflight prep and/or planning	323	10.2
Pilot–failed to obtain/maintain flying speed	329	10.4
Powerplant–failure for undetermined reasons	235	7.4 ⎫
Pilot–Mismanagement of fuel	229	7.2 ⎬ 20.3%
Fuel exhaustion	180	5.7 ⎭
Pilot–misjudged distance and speed	164	5.2
Pilot–improper level off	161	5.1
Pilot–selected unsuitable terrain	160	5.1
Material failure	154	4.9
Pilot–failed to maintain directional control	153	4.8
Pilot–continued VFR flight into adverse weather conditions	152	4.8

Table 7.2 Most prevalent types of accidents, all fixed wing aircraft, 1981 and 1976–1980. (Annual review of accident data, US general aviation 1976–1981.)

Type of accident	1981 No.	1981 %	1976–1980 Mean	1976–1980 %
Engine failure or malfunction	829	26.2	896.6	24.9
Collision with object	512	16.2	547.4	15.2
Ground/water loop-swerve	361	11.4	462.8	12.8
Hard landing	168	5.3	224.6	6.2
Collision with ground/water – controlled	180	5.7	178.4	5.0
Overshoot	162	5.1	165.4	4.6
Stall/mush	124	3.9	164.2	4.6
Nose over/down	95	3.0	127.4	3.5
Collision with ground/water – uncontrolled	120	3.8	125.8	3.5
Stall	121	3.8	120.8	3.4
Undershoot	108	3.4	107.8	3.0
Stall/spin	66	2.1	83.2	2.3
Wheels-up landing	49	1.5	69.4	1.9
Airframe failure in flight	49	1.5	57.6	1.6
Gear collapsed	48	1.5	47.6	1.3
Gear retracted	37	1.2	41.2	1.1
(All other types)	133	4.2	182.4	5.1
Total	3162	100.0	3602.6	100.0

Every year a large number of fuel-related accidents occur and in the UK during 1984 there were 40 instances of engine failure out of a total of 140 reportable accidents to fixed wing aircraft under 2,300 kg. Many of these engine failures were related to fuel exhaustion and fuel starvation, as the

Fig. 7.11 (Courtesy of the Australian *Aviation Safety Digest*)

FUEL MANAGEMENT

following brief summaries extracted from the annual survey of accidents published by the CAA show.

'The pilot noted that the fuel gauges were erratic but stabilised to half full after 10 minutes of flight. Because of this he decided to divert to Halfpenny Green but fog obliged him to continue to Bristol. During a radar-assisted approach the gauges dropped to empty, the engine stopped and a forced landing was made in a ploughed field, damaging the mainspar and landing gear ...'

'The aircraft ran out of fuel and made a forced landing, during which damage was sustained ...'

'The aircraft landed on a disused airfield, out of fuel. The left wingtip hit a traffic sign. Investigation found the fuel tank water drain open ...'

'The aircraft made a forced landing in a field two miles short of the runway when the engine ran roughly and then stopped. It had taken off six hours before with full long-range tanks for the purpose of aerial photography. The engine first ran roughly after five hours, this problem being cured by a reduction in rpm and application of carburettor heat. During the forced landing the nose wheel collapsed and the aircraft overturned ...'

'During a rolling take-off on a dual type conversion flight, the engine coughed and stopped at 200 ft. The instructor selected the rear tank, took control and turned left to avoid trees ahead. The aircraft banked sharply, struck the ground and cartwheeled. The forward fuel tank was empty and the gauge had repeatedly stuck at half full as the contents reduced. The engine stopped before it had time to draw fuel from the rear tank ...'

'Prior to a spraying flight which was expected to be of about three minutes duration, the fuel gauge showed the contents to be between one quarter full and empty. The low fuel light was not illuminated. On initial climb at about 70 ft the engine stopped and the aircraft was damaged during a forced landing. On impact the fuel warning light came on and the gauge read empty ...'

'Following the refitting of an overhauled engine and a test flight, the aircraft took off and proceeded normally for about 20 minutes. A sudden loss of power resulted in a forced landing; the aircraft undershot the selected field and went through a hedge. Prior to these flights the aircraft had not been flown for two years. The fuel system was heavily contaminated with foreign matter and a significant quantity of water. It had not been flushed out when the overhauled engine was refitted ...'

'On the initial climb the engine started to misfire. The instructor managed to restore power by pumping the throttle but at 500 ft the engine cut. Attempting a turn back to the aerodrome the aircraft crashed in the public road. After seven touch and go landings the aircraft had run out of fuel ...'

'Shortly after take-off the engine lost power and the aircraft forced-landed in the overshoot area. The wrong tank had been selected and the aircraft had run out of fuel ...'

'On the sixth flight, for the purpose of aerial application work, the engine stopped. During the forced landing the aircraft overran the selected field. The engine had stopped due to fuel exhaustion. A defective tank contents sensor unit is suspected ...'

'On a flight from Fenland to Southend the aircraft ran out of fuel. A forced landing was carried out in a field and the left undercarriage collapsed ...'

'Following the preflight inspection and power checks the aircraft took off on runway 26. At about 150 ft the engine surged and died, and the pilot made an emergency (landing with the engine surging at low rpm) on runway 25 which was just being cleared by an HS 125. The tower had thought the aircraft was going to hit either the car park or the terminal building.

Fuel was found to be pouring from a fuel drain. This was because the drains check had been done with the fuel selector in the OFF position, and the bayonet type drain valve had been left open. There is a warning in the Pilot's Operating Handbook on both these aspects ...'

'During the preflight inspection samples of fuel were taken from both tanks using the spring-loaded drains. There was no significant leakage and the reporter believes that there was not even a very slow drip following this operation. However, when the aircraft was established in cruise, fluid was seen to be leaving the trailing edge of the right-hand wing, so the left-hand tank was selected (as it contained more than twice the fuel required for the flight). After landing, the right-hand tank drain was leaking fuel in a continuous stream, 10–15 gallons having been lost in about 40 minutes ...'

'The pilot was making a long (6-mile) approach using the wing-down technique to combat drift. The engine started missing and picking up and he decided to land in a field below the approach path. He landed successfully without damage about $\frac{1}{2}$ mile from the runway.

No fault could be found with the engine and it is suspected that the cause of the misfiring was fuel starvation. Only two gallons of fuel, enough for 20 minutes' flying, were removed when the tanks were drained ...'

On some aircraft, which have a single ON/OFF selector valve for both tanks or when the 'both' position is selected it is common for one tank to feed down to low contents in advance of the other. Wingdown (or sideslipping) could uncover the feed pipe in the lower wing which may, in this case, have been the fuller tank. It is significant that in this incident, the engine picked up momentarily when the wings were levelled for touchdown.

FUEL MANAGEMENT

Reducing the risk of having a fuel-related accident

The majority of errors and assumptions which lead to fuel mismanagement can, in the main, be programmed out of your flight operations by paying serious attention to the following:

- Carefully read the aircraft manual in relation to the fuel system and how it is operated. Ensure you understand the information the aircraft manual contains with regard to the operation of the system, e.g. when to use the electric fuel pump, when to use mixture control and how to operate the auxiliary fuel system (when fitted). Do not assume the aircraft you will be flying has an identical fuel system to different models of the same types you may have flown many times before.
- Operate the fuel system intelligently and in accordance with the information contained in the specific aircraft manual.
- Understand the correct mixture control technique for the aircraft you are flying and use it whenever appropriate.
- Be aware that the fuel consumption tables as shown in the aircraft manual are based upon the use of 'lean' mixture.
- Ascertain your fuel consumption rate prior to flight and ensure you have sufficient fuel for the trip and a diversion, plus a reserve.
- Know the grade of fuel required by your aircraft and, whenever possible, personally monitor all refuelling operations.
- Know the usable quantity of fuel in your tanks before you take off.
- Ascertain that no contamination is present in your fuel and all fuel vents are unblocked and undamaged.
- Be aware of any common errors applicable to your fuel gauge indicating system. In this respect and when electrical fuel gauges are fitted it is good practice to note that the gauges are reading empty before switching on the master switch; fuel gauge needles have been known to stick in the full or partially full position. Making this check and watching for the fuel gauge needles to rise when the master switch is placed in the ON position will alert you to whether the fuel gauges are at fault in this respect.
- When fuel flow or fuel pressure gauges are fitted, monitor these when changing tanks during flight and when switching off electric fuel pumps.
- Know your mixture control procedure when changing from one tank to another.
- Double-check your selection lever/switch whenever changing tanks or operating any part of the fuel system.
- Never run a fuel tank nearly dry before switching to another one. This ensures you have some fuel left in the event of a 'frozen' selector valve or a malfunction (restriction of fuel flow) between the newly selected tank and the engine.

- Ensure all strainer valves have closed after checking for fuel contamination.
- Positively ensure that all fuel caps fit properly and are securely tight.

8
Weather considerations

Following the experience gained during training, many pilots become aware of their limitations and also those of the aircraft in relation to flight into adverse weather conditions. As a result of this they accept the importance of obtaining weather information prior to a flight.

However, in reviewing the accidents which occurred to general aviation fixed wing aeroplanes and helicopters in the UK for the 5-year period 1979 to 1983 we find that there were 39 accidents related to weather deterioration en route, which resulted in 75 people being killed and 10 seriously injured.

During the same period a total of 148 fatalities and 47 serious injuries occurred from all types of accidents. Thus for this period we see that over 50% of all fatalities were incurred in weather-related accidents. It should also be noted that the above number of weather-related accidents did not include those which occurred due to the adverse effects of wind, visibility, etc., during take-offs and landings.

From these figures it would appear that a number of pilots either ignore or pay little attention to the weather during preflight planning, or fail to plan properly or make the right decision at the right time, whether it be on the ground or in the air. Comments like these are only too easy to put down on paper, but the problems relating to pilots being caught in weather conditions beyond their ability are not so simple to resolve.

It is difficult to believe that most of these accidents were caused by the pilot 'pressing on regardless', and the clinical statements made under 'causal factors' in the accident surveys do little to reveal the underlying reasons. Such expressions as 'Continued flight into unfavourable weather' and 'Continued flight below minima' need examining to obtain an understanding of the real cause which led to the accident. In doing so we find that whereas an element of compulsion was present at the time there were also a number of uncertainty factors. For example, if at the preflight planning stage the cloud base was 200 feet, or the visibility 200 metres, the decision would have already been made for most pilots: a no-go situation. By far the largest number of occasions will, however, involve a situation where the weather is 'so, so', in that the cloud base could be 1,000 feet or a little more, with the visibility being rather poor, and with the forecast giving indications that the weather might improve followed by a front moving in, but timing rather vague, and so on.

The pilot is therefore left in two minds: on getting airborne it could turn into a reasonable flying day, or alternatively the weather could deteriorate and turn the flight into a traumatic experience. However, the pilot has arrived at the airfield, his passengers are also there and waiting, or his appointment at the end of the flight has already been made, so perhaps it would be worth having a try; after all, he can always turn back ... The die is cast, the application of common sense has been made ... one can always turn back. Armed with this thought the pilot completes his flight planning and eventually takes off with no intention of getting himself into a situation where his ability or inability to fly and navigate by sole reference to instruments will have to be tested.

What can happen next is something which has happened many times in the past. The weather he experiences does not occur in nicely defined levels of cloud base, visibility or precipitation, but rather as a somewhat murky grey hotchpotch of cloud with an indefinable base or patchy scud at varying levels. Occasional brighter patches may be seen ahead or to either side and for the moment it looks as though the weather could improve ...

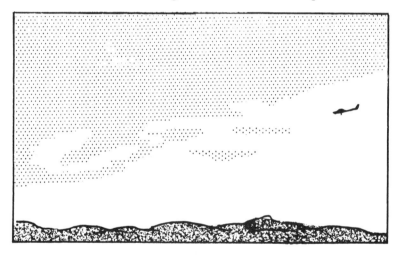

Fig. 8.1

... The urge to press on a little further is strong, after all it may not be long before the weather improves and, encouraged by this thought, the pilot continues. Visibility is poor and the cloud base begins to lower, so the pilot reduces altitude ...

In situations of this sort navigation becomes difficult and the largest part of the pilot's mental activity is taken up with attempting to obtain more frequent position checks than usual in order to maintain track. Now, whether this situation occurs shortly after take-off or in the middle of a flight which started in good weather is unlikely to affect what can come next – inadvertent penetration into cloud or conditions of extremely reduced visibility.

WEATHER CONSIDERATIONS

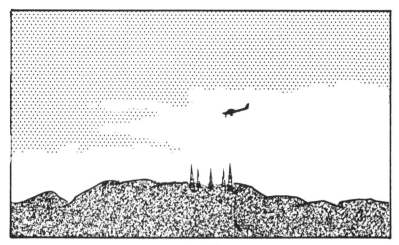

Fig. 8.2

Turning back at this stage is usually too late and it is at times like this that the pilot's ability to make clinical decisions is weakened by the thoughts uppermost in his mind: 'damn the weather! ... why did I get myself into this situation? ... I should have turned back before this ... will I be able to identify my position if I get clear of this muck? ... better go down to keep in sight of the surface ... must watch the instruments ... the altimeter ... must keep control of the aircraft ... better wait till I can see the ground before I turn round ... Christ! my wife and kids will never forgive me if I have an accident ... it's not getting any better ...' All these chaotic thoughts will be tumbling through his mind – as the aircraft makes contact with the surface.

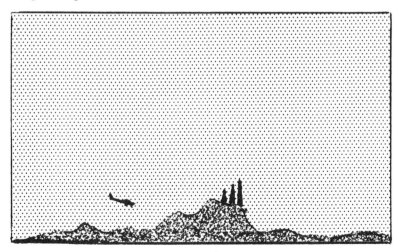

Fig. 8.3

So there we have it, just one more weather-related accident with its attendant fatalities. It was never planned, the situation just grew around the

Fig. 8.4

pilot's actions, and the weather; time and altitude were just not on his side, with inevitable results. With these facts in mind, what can be done to reduce the risk of this occurrence? Well, it all starts at the flight planning stage, and it is not all confined to the weather forecast, it also involves navigation planning.

Clearly, the weather information for the period of the flight must be properly interpreted and apart from understanding the forecast, one must also be able to determine how it will affect the following factors:

(1) Height of the cloud base en route.
(2) Visibility.
(3) Wind velocity.
(4) The possibility and amount of precipitation, including icing.

Put simply, these are the four main factors which sum up the practical aspects of meteorology for a pilot.

Height of the cloud base

Weather forecasts give cloud base in terms of height above sea-level, and therefore a forecast cloud base of 2,000 feet may be quite satisfactory for a flight over fenland areas but quite out of the question when contemplating a VMC flight over hilly or mountainous regions. So how does one determine whether a forecast cloud base is one in which a VMC flight can safely be made? This brings us to the sadly often ignored factor called 'safety altitude'. This altitude is calculated on the basis of having a certain minimum height above obstructions, and this includes high ground within a specified distance either side of the intended track and beyond the destination airfield.

To put the purpose of safety altitude into its proper perspective it must be understood that it is to cater for any deterioration in the weather during flight. Once calculated and entered in the flightplan/log it becomes a constant reminder to a pilot of how low he should descend en route in the event that the cloud base lowers or the visibility decreases. If one were to fly in a country which never has cloud and always has a visibility of 50 km or more then a safety altitude would be redundant (always bearing mind the separate issue of the Rules of the Air which regulate how low a pilot may fly over built-up areas, etc.).

In order to determine what constitutes a safe minimum altitude for any route it will be necessary to examine the elevation of the surface and any obstructions along and to either side of the planned route and just beyond the destination. Although there may not be a specific formula laid down in the regulations for flights conducted in accordance with the Visual Flight Rules, a sensible method would be to encompass the area 5 nm either side of the route and 5 nm beyond the destination aerodrome. Within this area, find the highest obstruction and then add approximately 1,000 feet. This altitude should be entered in the flight log and the pilot must discipline himself not to descend below it even if the weather deteriorates. This does not mean that he must plunge into any cloud encountered at this altitude but instead he must make the decision to divert from the route or turn back if the cloud base lowers to this altitude. Of course, once a pilot diverts from his planned route to proceed to an area where the cloud base is higher, he will need to calculate a fresh safety altitude and this will naturally take time – thus emphasising the need to divert early and not when the situation has reached a critical stage.

A safety altitude can only be effective if the pilot makes his decision at the right time and before getting into a situation where he has entered cloud. To emphasise this point, the accident summaries reveal that of the 39 weather-related accidents in the 5-year period mentioned earlier, 22 were caused by the aircraft simply being flown, apparently under control, into high ground varying from 600 feet to 2,800 feet. Whether the pilots concerned had calculated a safety altitude for their flights is not known, but if any of them did, for reasons unknown they did not abide by it. If they had, then over 50% of the fatalities caused by weather-related accidents would never have occurred.

Visibility

A reduction in in-flight visibility is one of the main reasons for pilots becoming uncertain of position or lost. This is particularly true when a pilot's ability to navigate has never gone beyond the simple process of map reading. The ability to read a map is only one of the essential ingredients required in order to navigate safely, and it therefore follows that pilots who learn to navigate properly by taking into account the passage of time between fixes, flying accurate headings and maintaining selected altitudes, will become lost far less easily than those who move a finger along a track line drawn on a topographical chart, and so spend most of their time looking from ground to the map or vice versa to seek assurance of their position.

One problem in relation to poor visibility is the training environment from which newly qualified pilots emerge. During training students are normally only authorised to undertake solo cross-country flights when the visibility will do little harm to the authorising instructor's ulcers, and this results in situations where students only fly cross-countries in visibilities of 10 km or more.

In the UK, however, once qualified, private pilots will have privileges which permit them to fly solo in uncontrolled airspace in minimum visibility of 1.5 nm at any altitude, and when carrying passengers to fly in a minimum visibility of 3 nm below 3,000 feet amsl and in 5 nm above 3,000 amsl. Most of the flying carried out by private pilots is conducted below 3,000 feet amsl, so this means that most flying in the private sector could legally take place in visibilities ranging from 1.5 nm to 3 nm.

To a VMC trained pilot flying at 120 knots in a flight visibility of 2 nm and using the finger-on-a-map technique this will present enormous problems in terms of workload and safety. For example, at an airspeed of 2 nm per minute he will only have 60 seconds to sight and identify a ground feature, and if he is flying downwind the time available will be even less and the problems increased. At the same time as these continuous 'seek and find' actions are being carried out, the aircraft will have to be controlled in relation to altitude, heading, lookout and the many other functions involved in operating the aircraft with safety. Added to which there are the problems involved when ground reference features are widely spaced, or conversely when a multitude of such features abound such as when overflying densely populated areas. In considering these facts it is easily appreciated how pilots can become disorientated, uncertain of their position and finally totally lost, when flying in poor visibility.

In the early days of flight, when controlled and regulated airspace barely existed, a pilot could flounder around, losing himself, finding his position and losing himself again, without being too much of a danger to other airspace users and, provided he stayed at a safe altitude, these circumstances could occur without undue danger to himself or his passengers. Today, however, the environment has changed and so have the relative speeds of aircraft, thus increasing substantially the dangers of attempting VMC operations in reduced visibility. It would therefore behove all pilots to appreciate these facts during their preflight planning as well as in the application of their in-flight decisions. If the reported or forecast visibility gives you concern then automatically it becomes a 'no-go' situation.

Although the use of 'safety altitude' primarily relates to a lowering cloud base its real purpose is to help the pilot maintain a safe altitude when his forward vision is obscured. The calculations made when considering safety altitude apply equally to forward visibility from the cockpit and to a lowering cloud base, and pilots should take this into consideration when contemplating flying at a lower altitude just because the visibility is worsening. There can be no argument against the fact that the closer to the surface a pilot flies the more difficult navigation will become because ground features are always more difficult to identify as one's line of sight becomes more oblique. There is also the added stress and worry brought about by the knowledge that the aircraft is much closer to ground obstructions at a time when forward visibility is reduced.

Thunderstorms and line squalls

Althought the majority of weather-related accidents are attributed to the presence of poor visibility or low cloud, a smaller but significant number occur because the pilot failed to appreciate the hazards of thunderstorms, line squalls or icing, either by paying too little attention to the possibility of their development during flight planning, or by allowing the flight to continue into icing conditions or becoming too close to dynamic weather disturbances.

Most general aviation aircraft are neither built nor equipped to fly in airframe icing conditions, nor to fly in close proximity to thunderstorms or line squalls. In the latter case the danger is compounded by the fact that thunderstorms often have a will of their own when it comes to their speed and direction of movement. For these reasons a separate list of advisory points covering these types of weather is given in the 'Reducing the risk' summary at the end of this chapter.

Airframe icing

When VFR flight can be conducted there will normally be no problem of ice forming on the aircraft unless flying through rain at temperatures below 0°C, or after descending rapidly into warm moist air after flying at levels where sub-zero temperatures exist. Prevention can therefore be ensured by avoiding flight through cloud or rain whenever the ambient temperature is at 0°C or below.

It will further be necessary to avoid parking an aircraft outside at night on occasions when the temperature is low and when sufficient humidity exists to produce hoar frost. If this or any other form of icing is present on a parked aircraft, it must be removed prior to flight since even the lightest formation of ice on the wings or control surfaces can significantly increase the take-off run or even prevent the aircraft becoming airborne.

Light single-engine aircraft rarely have any form of de-icing equipment, except a pitot heater. Larger aircraft are often fitted with some form of de-icing equipment which varies with the specific make and type. The equipment fitted will usually be one of three types – fluids, heatings systems, or rubber membranes. (*Note:* Information on the operation of de-icing systems will be found in the specific aircraft manual. However, an important rule for the use of the pitot heater is that it should be switched on prior to flight through rain, or when the outside air temperature is near to or below 0°C and flight through cloud is anticipated.)

Finally, bear in mind that most light aircraft are not equipped, nor cleared, for flight into known icing conditions and pilots should therefore avoid areas in which icing risks exist.

Weather problems start on the ground

Whilst in no way attempting to soften the general aviation pilot's responsibilities prior to, or during flight, one cannot leave the subject of weather-related accidents without writing a few words on the difficulties and problems experienced by many pilots in obtaining easily understood aviation weather reports or forecasts.

In this respect it must be borne in mind that at present over 60% of the total flying hours are carried out by the general aviation sector and over 60% of the world's pilots are private pilots. Thus the largest number of hours are done by pilots who are more often than not less experienced and less qualified than those who are employed in the commercial air transport sector, which is one where pilots work in a tightly controlled environment provided with excellent weather information facilities.

Additionally, the general aviation pilot, in the main, operates in the lower levels of the airspace, where the greatest potential for hazardous weather exists. It must further be recognised that whereas weather-related accidents are not very high on the list of causal factors, a general survey reveals that one-third of the fatal accidents have a weather-related factor. It cannot be denied that any difficulty in obtaining and interpreting weather information contributes, in no small way, to the occurrence of these fatal accidents.

Bearing in mind that this book is written about reducing the risks in aviation, it is pertinent to point out and highlight a situation which exists in some countries and which is scheduled to worsen in the future due to the introduction, or the tightening up, of financial methods to recover the costs of providing a weather service to aviators. On the subject of meteorological charges it is worthy of note that far less than half of one per cent of the total population in any aviation-oriented country are pilots, and whereas the general public in any country obtains its weather information service free of any charge, considerable pressures are now being applied by some governments and aviation authorities for pilots to pay for the weather information they need to operate an aircraft with safety – information which is certainly more vital than the need to know when to wear a mackintosh or carry an umbrella. If such governments or aviation authorities are prepared to put commercial pressures above the saving of human life then they cannot blame pilots who may be following their example and fail to obtain weather

WEATHER CONSIDERATIONS

information because of its associated cost. Fortunately, most countries have enlightened governments which ensure that this cost is met from the public purse and under normal conditions pilots are able to obtain the information required before a flight commences. It behoves every pilot to ensure not only that he obtains this information, but that he understands its implications and how they can affect the flight.

Good pilots abide by the laws of common sense and do not allow themselves to get into irretrievable situations due to weather deterioration. The philosophy of being firm over 'no-go' situations and the making of decisions to divert or turn back *at an early stage* when weather deterioration occurs cannot be overstated.

It is worth mentioning that the obtainment of a degree of instrument flying capability, or even an instrument rating, is in no way indicative that a pilot can operate in all types of weather.

The size and type of aircraft flown, the equipment it carries and the recency experience of the pilot is, in the case of a private pilot, largely dependent upon financial considerations. The weather, however, is totally outside the control of any person and therefore the measure of real pilot competence lies in his ability to assess the anticipated weather conditions and make a correct 'go or no-go' decision prior to flight. Additionally, he must always plan for an alternative course of action should the weather deteriorate to an extent which would make the continued safety of a flight uncertain.

Fig. 8.5

FLIGHT SAFETY IN GENERAL AVIATION

The following is an extract from a special study of general aviation weather-related fatal accidents published by the National Transportation Safety Board of the USA. It may be of interest to carry out a personal comparison check in relation to your own flight experience, qualifications and flying activities to see how you rate.

Statistical summary

Based on the statistics presented, a pilot most likely to have been in a fatal, weather-involved, general aviation accident:

(1) Received an adequate preflight weather briefing by telephone from a Flight Service Station which utilises weather service forecasts which were reasonably accurate.
(2) Was proposing a pleasure flight.
(3) Had between 100 and 229 flight hours.
(4) Had less than 100 hours in the type of aircraft.
(5) Had less than 50 hours in the 90 days before the accident.
(6) Had a private pilot's certificate.
(7) Did not have an instrument rating.
(8) Had no actual instrument time, but did have between 1 and 19 hours of simulated instrument time.
(9) Had not filed a flight plan.
(10) Was between the ages of 41 and 45.
(11) Crashed in IFR conditions, probably in fog or rain during daylight hours.
(12) Was accompanied by at least one passenger.

If you recognise yourself in the above summary, you can nevertheless quite easily opt out of becoming a weather-related fatal accident statistic by being very positive over your 'no-go' decisions prior to flight and by ensuring you make your decision to divert or turn back during flight at the right time.

Regardless of the privileges awarded by an instrument qualification the pilot must learn to assess his own capabilities and these will inevitably vary in accordance with recency practice in instrument flight. He must certainly not hesitate to restrict his flight operations to match his current skill, the type of aircraft flown, and the instrumentation and radio navigation equipment carried in the aircraft, and that available at the planned destination and alternative aerodromes.

The weather conditions which prevail or which are anticipated from the time of take-off, and during the en-route and arrival phases, are vitally important to the pilot's determination of whether the flight can be safely conducted. A thorough understanding of the methods whereby meteorological information is made available to the pilot, and his ability to interpret and assess this information, is therefore a positive requirement, and particularly so for the pilot who wishes to exercise the privileges of an instrument rating.

Tables 8.1, 8.2 and 8.3, taken from published accident data, give a clear indication of how 'adverse weather' appears as a causal factor in general aviation accidents. Table 8.1 reveals the percentage of fatal accidents from this cause and emphasises the high degree of risk involved.

Table 8.1 Broad cause/factor assignments – all accidents. All operations, 1981 and 1976–1980. (Annual review of aircraft accident data, US general aviation, 1981.)

Broad cause/factor	1981		1976–1980	
	No.	%	Mean	%
Pilot	2768	79.0	3207.4	81.2
Terrain	619	17.7	925.6	23.4
Weather	947	27.0 ←	852.6	21.6 ←
Powerplant	590	16.8	579.8	14.7
Personnel	311	8.9	366.2	9.3
Airport/airways and facilities	299	8.5	293.8	7.4
Miscellaneous	188	5.4	148.2	3.8
Landing gear	123	3.5	143.6	3.6
Undetermined	79	2.3	86.0	2.2
Systems	57	1.6	55.0	1.4
Airframe	56	1.6	47.0	1.2
Rotorcraft	40	1.1	41.0	1.0
Instrument/equipment and accessories	12	0.3	22.0	0.6

Note: Whilst the number of accidents dropped by 11.3% from the mean for the five-year period 1976–1980, the number of accidents in which the weather was a cause/factor increased by 11.1%.

Table 8.2 Broad cause/factor assignments – fatal accidents. All operations, 1981 and 1976–1980. (Annual review of aircraft accident data, US general aviation, 1981.)

Broad cause/factor	1981		1976–1980	
	No.	%	Mean	%
Pilot	568	86.9	575.4	87.0
Weather	267	40.8 ←	257.6	39.0 ←
Terrain	48	7.3	91.8	13.9
Personnel	64	9.8	73.6	11.1
Powerplant	47	7.2	55.2	8.3
Undetermined	50	7.6	50.2	7.6
Airframe	40	6.1	21.2	3.2
Miscellaneous	29	4.4	20.4	3.1
Airport/airways/facilities	9	1.4	8.8	1.3
Rotorcraft	7	1.1	8.2	1.2
Systems	10	1.5	7.8	1.2
Instrument/equipment and accessories	4	0.6	5.8	0.9
Landing Gear	0	0.0	0.6	0.1

Table 8.2 lists the number and percentage of fatal accidents in which each broad cause/factor was cited in 1981 and in the base period 1976–1980. The percentages of fatal accidents with pilot and weather cause/factors changed little in 1981 when compared to the prior 5-year period.

Table 8.3 Survey of accidents to Australian civil aircraft, 1981

	Commuter	Charter	Agriculture	Flying training	Other aerial work	Private/ business	Total
Fatal accidents, broad factors	1	14	15	2	22	51	105
Pilot		9 64%	8 53%	2 100%	11 50%	27 53%	57 54%
Other personnel		4 29%	2 13%	2 100%	9 41%	5 10%	22 21%
Airframe			1 7%	1 50%		3 6%	5 5%
Powerplant		3 21%	3 20%		2 9%	3 6%	11 10%
Systems		1 7%		1 50%			2 2%
Instruments, equipment, accessories			1 7%		1 5%		2 2%
Rotorcraft					1 5%		1 1%
Airport/airways facilities					1 5%	1 2%	2 2%
Weather		4 29%			3 14%	17 33%	24 23%
Miscellaneous	1 100%	3 21%	3 20%		6 27%	22 43%	35 33%
Total	1	24	18	6	34	78	161

Note: Percentages represent the frequency of occurrences of each broad factor in relation to the total accidents in each kind of flying. As there may be more than one of the above broad factors listed in each accident, it follows that the percentage total for each kind of flying will exceed 100.

In 1983 the results of a special study in relation to loss of control accidents in 'adverse weather' were published in the FAA Airworthiness ALERTS. The study was confined to flying activities within the United States, but the following extract from this study should be read by all pilots.

The data obtained from the Accident/Incident Data Systems (AIDS) indicate the number of loss of control accidents in adverse weather has increased from 67 in 1977 to a peak of 90 in 1979, decreased to 72 in 1980, and then increased to 77 in 1981. The number of fatalities resulting from these accidents has increased from 108 in 1977 to a peak of 187 in 1979, then decreased to 157 in 1980, and increased to 164 in 1981. The number of estimated hours flown, which was obtained from the General Aviation Activity and Avionics Survey, dated 1981, was 35.8 million in 1977, increased to 43.3 million in 1979, then decreased to 41 million in 1980, and further decreased to 40.7 million in 1981.

The total number of loss of control accidents in adverse weather during the 5-year study period was 387 and the total number of fatalities was 796, resulting in a fatality rate of 2.05, which ranks number one in fatalities per accident. Of the total number of accidents, 342 (88%) were fatal and 291 (75%) were in instrument meteorological conditions. Only 136 (35%) of the pilots involved in these accidents were instrument rated.

Loss of control accidents in adverse weather have a direct relation to seasonal weather changes. The high accident periods are in April and November.

The study reveals that complex high performance aircraft are more likely to be involved in this type of accident. The table shows the most involved aircraft rates based on aircraft population. Of the first four aircraft with the highest accident ratio, three are complex, high performance aircraft. This table was included in the special study to show that the types of aircraft most involved in these accidents are faster, longer-range aircraft, and not to imply that one aircraft is any safer than another.

Aircraft most involved in this type of accident

Rank	Make	Aircraft group	No. accidents	No. aircraft	Accident rate
1	Beech	36	10	1869	0.006
2	Cessna	210	29	6576	0.004
3	Gulfstream	AA1,AA5	12	2986	0.004
4	Beech	35	26	7117	0.003
5	Piper	PA28	61	23094	0.002
6	Mooney	M20	12	5990	0.002
7	Cessna	172	35	25899	0.001
8	Cessna	182	17	14254	0.001
9	Cessna	150	22	20682	0.001

Extracts from the CAA publication *Accidents to Aircraft on the British Register* concerning general aviation accidents with weather as a causal factor:

'The pilot was en route from N Ireland to the Isle of Man and in radio contact with Ronaldsway Approach. He was 3.4 miles DME from the VOR cleared to land. There was no further RT contact with the aircraft. Search and rescue action was initiated. The following morning the wreckage was discovered at 1,150 ft in the south face of high ground which rises 1,434 ft ...'

'The aircraft was flying from Carlisle to Little Snoring on a VMC flight. The pilot later changed his route and was seen flying in and out of cloud. The aircraft crashed into a near vertical rock face. It is hypothesised that the pilot mistook Workington for Whitehaven, turned onto his original track and flew along Ennerdale...'

'On a flight from Edinburgh to Glenrothes an ATC clearance was acknowledged as routeing via the Forth Bridges. Special VFR not above 1,500 ft. The cloud base was 800 ft at Edinburgh but another pilot warned that it was down to 300 to 400 ft further north. A witness sighted the aircraft about 300 ft above ground clear of cloud. Later, a further witness heard the aircraft fly overhead in cloud followed by the sound of its impact. It had struck the hill 15 ft below the highest point at an altitude of 770 ft amsl. The procedure and limitations concerning special VFR clearance in the Edinburgh Control Zone are clearly documented in the UK Air Pilot...'

'En route to Thruxton an unforecast deterioration in the weather occurred giving fog, low stratus and some rain. Believing this was only local, the pilot continued but found himself above extensive low stratus. In a clear patch he descended to land in a field which turned out to be wet standing corn. The aircraft nosed over...'

The pilot ascertained the actual weather for destination and en-route aerodromes 1 hour before take-off. During the flight, conditions became poor north of Lewes and worse near the coast. It was decided to return to Horsham, but the weather deteriorated and a precautionary landing was carried out. During the final stages of approach the aircraft rolled and the nose pitched down. Corrective action was insufficient to prevent a heavy landing...'

'The aircraft was returning at night to Manchester, cleared to enter controlled airspace, Special VFR, not above 3,000 ft. During its approach pattern it passed on the lee side of Kinder Scout, the highest point of the Derbyshire Peak District, elevation of 2,088 ft. The pilot reported that he was in a severe downdraught and was unable to maintain height; thereupon the controller instructed him to resume his own navigation to keep clear of high ground. The pilot soon advised ATC that he was 'going down at 1,000 fpm' and declared an emergency. Despite the use of full power and best rate of climb speed he was unable to maintain height as he encountered a strong downdraught in smooth air on the leeside of the hills. On being carried down through 1,840 ft the air became extremely turbulent. He was aware of the loom of hills ahead and was able to reduce airspeed to 50 to 55 knots before striking a steep slope at 1,650 amsl. The aircraft turned over on impact but both occupants were only severely bruised. Met report gave airport surface wind as 280/5 knots but radar recording of aircraft track gave 3,000 ft wind at accident site, 290/45 knots...'

'The aircraft was inbound on a VFR flight plan and coasted inbound at Dover at 1557 hours. By 1613 hours Biggin Hill weather had started to deteriorate from the south. The pilot informed ATC that his ETA was 1630 hours and was advised of the weather conditions. At 1625 hours he was informed that fog from the south of the airfield had reduced visibility to 500 metres. At 1628 hours ATC informed the pilot that the control tower was in fog. The pilot requested the QFE and asked for it to be repeated. Subsequent calls from ATC were not answered. The aircraft crashed 1.25 nm to the east of the airfield, having struck treetops and descended into high tension cables ...'

'The VRF flight plan was filed for a night route from Inverness to Glasgow. Following a descent from FL 70 to 3,500 ft the aircraft crashed into the eastern face of Ben Ledi some 200 ft below the summit of 2,870 ft. Severe icing and turbulence were present at the time of the accident and a considerable portion of the descent was conducted in IMC and in a cold, soaked aircraft ...'

'The aircraft was engaged on a night navigation exercise on a Special VFR clearance. ATC advised the pilot of his terrain safe altitude. The pilot descended below this safety height so as to keep clear of cloud. After several two-way transmissions, radio contact with the aircraft was lost. The wreckage was found the next morning on gently rising ground close to the summit of the moor ...'

'The pilot booked out for a VFR flight from Blackpool to Bournemouth although advised of deteriorating weather en route. The aircraft could not be traced on radar and when the pilot was lost in the Welsh hills, he was advised to turn back. The aircraft crashed at 1,200 ft under power in poor weather ...'

'The pilot made a go-around from his first approach, having arrived at the airfield in patchy stratus. The second approach resulted in the aircraft touching down some two-thirds down the runway. Heavy braking on short, wet grass caused the aircraft to skid along the remaining runway and it slid off the end, hitting a gate and fence ...'

'The aircraft, on a local flight, diverted to Nottingham due to adverse weather and made an approach with a strong crosswind. At 100 ft turbulence and heavy rain were encountered but the pilot elected to land since a missed approach would have put the aircraft in IMC and he was not instrument rated. The aircraft bounced twice and drifted off the runway, finally bouncing on the grass which caused the nosegear to collapse rearwards ...'

'As the weather deteriorated en route, the pilot decided to turn back. It grew worse and a precautionary landing was effected. The ground, softened by rain, tore off the nose wheel on a ridge ...'

'En route from Sherburn-in-Elmet to Dundee the pilot encountered a lowering cloud base and falling visibility. He chose a field for a precautionary landing but during the landing run a gust lifted the right wing and the aircraft came to rest on its nose ...'

'The pilot was accompanied by an experienced flying instructor. The flight was planned as a VFR flight, but the crew encountered 7 octas low cloud and attempted a descent through a gap in the cloud cover. The aircraft entered cloud and collided with a TV aerial mast supporting cable. The aircraft fell on to a concrete block house – all occupants of the aircraft were killed ...'

'The aircraft was seen to enter an isolated but active shower cloud, producing hail on the ground. It was seen to emerge from the cloud base in a steep spiral with most of the right wing missing. After a short delay large pieces of structure fell from the cloud ...'

'A precautionary landing became necessary due to unforecast deteriorating weather. Due to a down slope in what appeared to be a suitable field the aircraft bounced on first touchdown and because of the soft ground, nosed over during its second touchdown causing substantial airframe damage ...'

'On a flight from Ledbury to Perranporth the weather started to deteriorate near Launceston. The pilot decided to make a precautionary landing in a field. Because of the proximity of power cables he was unable to land into the wind. Wet grass and poor braking action resulted in the aircraft striking the hedge at the end of the landing run ...'

'In deteriorating weather the pilot carried out a precautionary landing in a field of corn. The aircraft nosed over in soft ground ...'

'The aircraft attempted to take off in a sleet shower and strong wind. On rotation an uncontrollable roll to the right developed and the right wingtip contacted the ground ...'

'The aircraft flew from Stebbing to Stapleford in VMC. On the return flight the weather deteriorated with cloud lowering to 200 ft and poor visibility. The pilot, who did not have an instrument rating, declared an emergency and made a precautionary landing in a field. Touchdown was normal but the nose wheel dug into soft ground causing damage to the propeller and collapsing the noseleg ...'

The student pilot was on a cross-country exercise at 800 ft agl when he encountered poor visibility conditions and reported he was unsure of his position. He was advised to climb to 3,100 ft, the minimum safe altitude for the area. The aircraft disappeared from the radar screen and was found crashed at 540 ft amsl ...'

Reducing the risk of accidents due to weather

Bering the previous facts in mind your risk factor will be significantly reduced provided you:

- Obtain proper weather information prior to flight. When reading forecasts and reports, visualise the weather being described and the effect upon the cloud and visibility en route.

WEATHER CONSIDERATIONS

- Relate the anticipated or known weather conditions to the type of terrain you will be overflying.
- Pay particular attention to the time period of the forecast and relate it to your estimated time of departure and arrival plus a good allowance for any delays before take-off, or possible diversions en route.
- If a delay occurs to your departure time, reconsider the weather situation.
- Study your chart for high ground and obstructions at least within 5 nm either side of the track and the same distance beyond your intended destination.
  Note the relevant heights and add 1,000 feet to arrive at a minimum safe altitude to fly en route.
  Enter this safety altitude in your flight log so that it is easily read at any stage of the flight.
- During flight never fly below the safety altitude you have calculated unless you have your destination in sight and have established that it will remain in sight throughout the arrival phase.
- If the cloud base lowers to your safety altitude whilst en route, do not assume that this will only be a temporary condition – the odds are that it will become lower and often that the forward visibility will often deteriorate as this occurs.
- Have your diversion aerodrome(s), together with their radio frequencies, noted in your flight log and implement a diversion decision should the cloud base lower to your safety altitude. A delay in arriving at your original destination is by far preferable to the unknown hazards implicit in the 'press on regardless' approach to cross-country flying.
- If you are instrument qualified in current instrument flying and radio navigation practice, bear in mind the following points before deciding that you can tackle the cloud en route whilst maintaining your safety altitude:
  (a) The freezing level – remember that very few general aviation aircraft are certificated for flight into known icing conditions.
  (b) The possibility of embedded cumulo-nimbus or turbulence in the cloud being penetrated.
  (c) The radio navigation facilities available to you – both in the aircraft and on the ground en route.
  (d) The instrument radio let down facilities available at your destination or suitable alternate(s) in case you have to divert when flying in instrument conditions.
  (e) Are you in current practice at using the available let down aids. If not, then use them with caution, or don't use them at all.

  If you are not completely competent and confident that you can complete the flight safely after reviewing the items in (a) to (e) you can make only one sensible decision – to remain in VMC or divert to an alternate, or return to the departure aerodrome. To add emphasis to the need to make the correct decision, it should be borne in mind that one-third of all fatal accidents are weather-related.

Avoidance of flight through or near thunderstorms

Although the mechanism involved in the formation of a thunderstorm is very complex, the advice to pilots regarding flight through, or in the immediate vicinity of, these violent forms of weather is not, and the simplest and safest policy for all pilots of general aviation aircraft can be summed up as follows:

- Avoid flying underneath a thunderstorm area, even though you can see through to the other side. The lower levels of the cloud and that region immediately below will be an area of intense turbulence.
- Keep well clear of any form of active thunderstorm. A good rule of thumb when deciding how close a light aircraft can be safely flown to such an area is, 'allow one mile clearance for every 2,000 ft of its vertical development'.
- Never attempt to fly through or below a line squall. These intense weather areas may produce violent turbulence, and lightning, and can extend along a line of considerable length. They can therefore be difficult to circumnavigate, and it is usually advisable to turn back or land at the nearest available aerodrome until it is safe to continue. Remember, line squalls move rapidly and sweep across an area quickly.
- Never assume that thunderstorms move in the direction of the general wind. Thunderstorms often generate their own winds and on occasions have been known to move in precisely the *opposite* direction to the general wind prevailing at the time.
- Never assume that any thunderstorm will be of light intensity. They can reach the mature stage with remarkable speed, and an incorrect decision can leave a pilot trapped inside an area of intense and violent weather activity.
- Never attempt to climb over the top of large and rapidly building cumulus cloud. Such formations often rise more quickly than the maximum rate of climb of a small aircraft.

Reports of lightning strikes to aircraft are fairly rare, but they can puncture the skin of the airframe and cause damage. A more common hazard is that lightning flashes occurring close to an aircraft can temporarily blind the crew and this could create a serious situation. Additionally, lightning discharges can disrupt radio communications and make certain radio navigation equipment unreliable. Further to this, intense static can be encountered which will also reduce the effectiveness of radio equipment.

Turbulence within and around the vicinity of a thunderstorm will normally be of such intensity that it precludes safe penetration, or flight close to, by light aircraft even when flown by skilled instrument-rated pilots. The violence of the updraughts and downdraughts can easily exceed the load factors which small aircraft are designed to withstand.

If the top of a thunderstorm exceeds the tropopause level, the storm is extremely hazardous and must be carefully circumnavigated at a distance no less than 10 miles from its nearest point. If tops exceed the freezing level by more than 10,000 feet, expect hail.

Finally:
- Aviation weather forecasters issue warnings of active thunderstorm areas in routine forecasts and in Sigmet messages: therefore, study forecasts and take note of them.
- If you find unforecast thunderstorm activity occurring during flight, do report this to the ATC services as soon as possible.
- When operating in areas of strong turbulence or while circumnavigating thunderstorm activity, do ensure that loose articles are firmly stowed and all safety belts or harnesses are tightly secured.
- Do remember that intense turbulence can be experienced many miles away from the location of a storm: this distance will usually be greatest when a line of thunderstorms is developing. Therefore when taking off from aerodromes which apparently lie well ahead of the storm activity you will need to exercise special caution and anticipate strong gusts and wind shear effects.
- Always avoid flying below the overhang area of cloud, ahead of the storm region, as it may contain large hailstones.
- Do remember that large altimeter errors may occur in the immediate vicinity of thunderstorms and not just inside them.
- Years of studying accident reports show that a storm moving across the ground at 20 knots is very hazardous, with moderate to severe turbulence, gusts to 50 knots, and horizontal wind shears of 25 knots. One moving over the ground at 30 knots will have within it severe to extreme turbulence, gusts to 60 knots, and horizontal wind shears of 30 knots.

Fig. 8.6

Although an epitaph comes to us all in the end we can at least do everything possible to avoid it reading like this...

Continued VFR flight ... into adverse weather!

9
The influence of pilot distractions on aircraft accidents

In reviewing the various causal factors which are accident-inducing it becomes clear that pilot distractions are high up on the list. The USA National Transportation Safety Board statistics reveal that most stall/spin accidents occur when the pilot's attention is diverted from the primary task of flying the aircraft. During a special survey, it was revealed that 60% of stall/spin accidents occurred during take-off or landing and 20% were preceded by engine failure (a most significant distraction). Other distractions included preoccupation inside or outside the cockpit, whilst changing power, configuration or trim, manoeuvring to avoid other traffic and whilst clearing hazardous obstacles during take-off or landing.

Causal factors

For the purpose of discussing distractions and the way they influence accidents, in the following paragraphs they can be divided into two classes:

(1) Those which are actually created by the pilot due to lack of thought, or lack of proper planning.
(2) Those which occur as a result of other influences, such as unexpected deterioration in weather, equipment unserviceability, or sudden emergency, etc.

The first of these two classes can be significantly reduced as they are entirely pilot-sponsored. They are the result of not planning a reasonable spread of the cockpit workloads in relation to time, thus resulting in a number of cockpit activities crowding up on each other.

During flight there will, at any time, be a need for physical and mental activity, and both of these will be taking place simultaneously. The amount of activity will depend upon the type of flight, e.g. a relatively simple recreational flight in the local area, or a navigation flight along an unfamiliar route to an aerodrome not previously visited. In the case of a deterioration in the weather or the occurrence of a minor unserviceability when flying in an area relatively close to the base aerodrome, the pilot's workload will be considerably less than if the same circumstances occur in the middle of a navigation flight involving some distance over unfamiliar territory. In the latter case the pilot will be engaged in a high degree of mental activity to ensure the navigation task is being conducted properly.

If at any time the pilot feels uncertain of the aircraft position, his mental activity increases and creates a distracting element which can interfere with his flying the aircraft. If at the same time an equipment malfunction occurs, a further distraction is added, and so on.

Planning the cockpit workload

When planning prior to a flight, therefore, an important job is to consider the number of tasks known to be required during the flight and ensure as far as possible that they are spread properly so that the pilot does not find himself in a situation where he is having to do multiple tasks at the same time. This is, of course, easier to say than do, but on the other hand it is is not too difficult to bend one's mind to the task of mentally visualising the various cockpit actions so that they become more of a continuous flow than sudden bursts of activity.

For example, during a navigation trip it is totally unnecessary to continously occupy oneself in identifying ground features or checking position by reference to radio navigation aids. This action, depending upon the type of route to be flown, can normally be adequately covered by making position checks every 5, 10, or even 15 minutes – thus promoting a more relaxed atmosphere in the cockpit and giving reasonable time between position checks to attend to other routine matters such as noting fuel contents, consumption rates, engine and equipment readings, etc. Furthermore, it allows time for the necessary mental activity to take place in a calmer atmosphere, more conducive to making qualified judgements or decisions, should the unexpected occur.

Organising this one facet of the cockpit workload in a sensible fashion allows the pilot to remain in a state of mental preparedness and affords him the opportunity, for example, of listening out to weather advisories en route, and similar activities.

Distractions leading to loss of control

To emphasise the part that distractions play in the chain of events which lead up to accidents, one has only to look at the figures of accidents listed under the heading 'loss of control in flight'. Most of these occur during the approach and landing phase, or during go-arounds, or the initial climb after take-off.

Loss of control in flight usually means that the pilot allowed the airspeed to get so low that he was unable to maintain effective control over the aircraft; this means a situation where the aircraft was at the stall or close to it. However, it was not the aircraft that put itself into this situation, it was the pilot who, through lack of attention, failed to monitor the airspeed and thus created the unsafe flight condition. Why? ...

Clearly, all pilots are made aware of the importance of maintaining a safe airspeed at all times and particularly when they are near the surface, and therefore in most cases it was probably due to the pilot allowing himself to become overloaded at a critical stage of the flight.

Before leaving the subject of self-induced workloads it would be appropriate to refer to the comments made earlier in this book concerning the importance of habits.

A typical example of pilot-induced distractions

In this respect pilots should be aware that any break in the routine flow of his checks or procedures can often be the equivalent of a distraction. The following extract is a report made in a NASA Safety Bulletin, which demonstrates how easily this can happen.

In the Nick of Time

'... I have read many reports of near misses and other such episodes; this one is a near gear up landing... It took me quite some time to figure out the whys and wherefores. In a nutshell, I was preparing an applicant for a type rating. I gave him a simulated wing/wheel well overheat, requiring the gear to be lowered. The objective was to prepare him for a single-engine ILS. He did exactly what he should have done: he reached for the gear handle. Since we were eight miles from the VOR and he was to hold, I said, "OK – leave the gear till later" (Mistake No. 1). He said, "OK". In his mind the gear was down (No. 2). I was expecting him to lower the gear at Glide Slope intercept (No. 3). He obviously wasn't, since in his mind it was already down. There was another aircraft doing ILSs, so we were distracted (No. 4). At GS intercept he did not put the gear down (No. 5) and I didn't check, as I was trying to locate the other aircraft already on the missed approach (No. 6). At GUMPS time (half a mile out) I didn't check because of concern for a NORDO (No. 7), and luckily I have been doing "flare checks". Just before I commit myself I have been doing a "three in the green, no red, hydraulics OK, cleared to land" check. That is why we did not have a gear up accident, but why we did scrape the tail skid. He did use our check-list but *in his mind* the gear was down. My excuse was distraction. Needless to say, my procedures have changed. Needless to say, he now also has a flare check over and above GUMPS ... Moral: after GUMPS, still do a FLARE CHECK! It saved two licences and one airplane...'

Reducing the risk of distraction-type accidents

In relation to distractions which are not directly induced by the pilot's actions it is clear that the events leading up to such circumstances are mainly caused by malfunctions of the aircraft controls, equipment and engine(s), or by problems associated with passengers and other factors which occur unexpectedly.

The degree of distraction created by engine or equipment malfunction can be appreciably reduced by the pilot having a sound knowledge of the recommended procedures which are prescribed in the aircraft manual. Although these procedures are normally learnt and applied during practice at the initial stages of a pilot's training, they are often sadly neglected after the pilot has qualified for a licence. The moral is to place plenty of emphasis upon recurrent practice of these procedures, thus:

- From time to time re-read the emergency procedures in the aircraft manual for the type you are currently flying.
- Whenever possible, mentally assume various malfunctions have occurred during flight and simulate practice of the correct procedure.
- Whenever practical, engage the services of an instructor or a check pilot to monitor your current skills and procedures.
- During your flight planning operations make sure you are planning to avoid problems and not just going through a routine in order to merely tick off items from your preflight check-list.

Fig. 9.1

10
Prevention is better than cure

Although fuel mismanagement is the cause of most 'pilot induced' engine failures, there are a number of other causes of a powerplant malfunctioning or ceasing to operate entirely. However, whenever a single engine aeroplane suffers engine failure there is usually only one consequence and that is a forced landing. This inevitably is a high risk situation whether the engine fails at high altitude in good weather or whether it fails when the aircraft is near the surface in any weather.

The basic ingredients which affect the success or otherwise of coping with this high risk situation will be:

The time available.
The weather conditions.
Whether a clear area on the surface is close enough and sufficiently large for the accomplishment of a safe landing.
The degree of competence shown by the pilot.

The first three of the above variables will be largely outside the control of the pilot and will be dependent upon the circumstances at the time, The last item, pilot competence, will depend not only upon aircraft handling skills but also upon the pilot being able to demonstrate good judgement and quick and correct decisions. Some points in relation to judgements and decisions during forced landing procedures are covered at the end of this chapter.

First, though, we shall deal with the question of why the powerplant malfunctioned and how it may have been prevented from failing; and secondly, how it may have been brought back into operation if the cause of failure was not mechanical or due to fuel exhaustion.

Aircraft maintenance

Dealing first with prevention and failures of a mechanical nature, e.g. cracked pistons, broken oil or fuel lines, malfunctioning fuel pumps or carburettors, etc., it is clear that the owner pilot will have greater control over the incidence of these occurrences than the pilot who hires aircraft from an aircraft operator. An aircraft owner should ensure that his aircraft is well maintained, has suitable hangarage whenever possible, and that he complies in all respects with advisory and mandatory servicing bulletins.

The owner pilot also has greater control over the way the aircraft is operated and he should take care to abide by the manufacturers' recommendations in relation to type of oil used, oil change periods, correct leaning procedures, power settings and restrictions, etc. Engine manufacturers publish considerable information with reference to how their products should be operated and all pilots should make every attempt to get hold of these publications because they contain a large number of useful hints and tips about engine operation.

The preflight inspection

The non-owner pilot, on the other hand, has to rely upon the integrity of the hiring company when it comes to regular maintenance and the implementation of work required by advisory and mandatory bulletins; he also has to accept that the aircraft has been flown by many other pilots who may not have been so careful in the way they followed the manufacturers' recommendations, if indeed they were always aware of them. Nevertheless any pilot hiring an aircraft has the responsibility of ensuring that the aircraft is within its maintenance check periods and that it is fit to fly. Apart from checking the aircraft documentation, pilots must, in order to meet their responsibilities, conduct an inspection of the aircraft before accepting it for flight. This preflight inspection is of vital importance to the safety of the pilot and his passengers, but alas, events have shown that only too often insufficient care has been used when carrying it out.

Just to illustrate this comment and to show the reader how some pilots fall short in their ability to conduct this type of inspection, an extract from a UK CAA General Aviation Safety Information leaflet is reproduced below. Whilst reading it, just bear in mind that early warning of a number of cases of mechanical failure through fatigue or other reasons might well have been discovered by the pilot if he had carried out a more careful preflight inspection.

The UK CAA Safety Data and Analysis Unit conducted an experiment on two occasions at the biennial Business and Light Aviation Show at Cranfield Airfield, England. The aircraft used for this experiment was a typical light general aviation type which was deliberately rigged with 11 defects. Pilots were invited to carry out a preflight inspection of the aircraft and complete a pro forma noting the number of defects they discovered. The following extract from the final report makes interesting reading and should give pilots a greater incentive to carry out more thorough inspections of this nature in their future flying activities.

CAA Report
'As some readers may know, the CAA arranged at the Cranfield Business and Light Aviation show for a College of Aeronautics Beagle Pup to be rigged with eleven defects. Throughout the show, people were invited to make an external preflight check to see if they could find the defects. The doors, cowling and fuel filler caps were sealed up. The

aircraft was parked on short grass and the weather throughout was dry and sunny, which gave ideal conditions since participants could take as long as they liked. One group of two commercial pilots and an operations officer found all but one of the defects, but they took 40 minutes. On the answer sheet we asked for the participants' qualifications. These included a Concorde pilot who owned a light aircraft, student pilots with only two hours, and a microlight pilot who classified himself as 'other'. Although the aircraft was in marvellous condition, having just been resprayed, many people found minor or suspected defects that were not part of the competition, these ranging from the date stamp on the fire extinguisher to the wing main spar being cracked. In fact the actual defects with the percentage of people who found them are as shown in the table.

Answers to faults on Beagle Pup

Item	Fault	Number of cases in 5 years to GA aircraft	Percentage entrants who found fault
1	Magnetos are live	4	91
2	Left hand aileron bonding wire disconnected	14 (cases of lightning damage)	98
3	Blocked pitot tube	17	88
4	Brake pipe wet with hydraulic fluid	21 (brake failure)	37
5	Exhaust stub missing	23 (Exhaust leaks)	12
6	Tow bar attached to nose gear	1	53
7	Creep marks not aligned on nose wheel	2 (valves failures)	76
8	Fuel tank contaminated with water	14	42
9	Tyre badly worn	45 (burst tyres)	71
10	Nut missing from right-hand elevator hinge	2 (hinge bolt missing)	67
11	Locking wire missing from rudder actuating arm	12 (loss of rudder control)	33

When the answer sheet was given to participants there were exclamations of disbelief that certain items had been missed, and remarks such as 'it makes you think' were common. Only one person (a PPL) found all 11 defects, six entrants found 10 of them (a commercial pilot, a private pilot, an engineer and three groups of two or three people working together). Fifteen people found 9 and thirty-six found 8.

The results by qualification were as one would hope:

Instructors	averaged 7.5
Engineers	averaged 7.3
Commercial pilots	averaged 7.2
Private pilots	averaged 6.8
Student pilots	averaged 6.2
Others	averaged 5.5

The things that people failed to find are worthy of comment. The brake unit soaked in oil was hard to spot, just as it would be in reality. The missing exhaust pipe was not spotted by 90% of participants, including some who were Pup owners. Perhaps it requires some knowledge of the aircraft, but many pilots fly more than one aircraft type or model and may not remember how many pipes there should be. The large quantity of water in the right-hand tank (a potential killer) was missed by over half the participants. They may have been put off by the fact that we sealed the tank filler caps, but we had placed a couple of convenient bottles near the nose wheel (some thought these were deliberately placed rubbish – perhaps we should have labelled the bottles). Some people drained fluid from the contaminated tank onto the grass (which tells you very little), others drained it into a bottle, found it smelt slightly of petrol and thought it was OK, even though it lacked the blue colour of AVGAS, and some drained the other tank, found petrol and did not try the contaminated tank. The elevator hinge nut was missed by nearly one-third of participants. It was nearest the fuselage and always in shadow and was about the last control surface attachment to be inspected. The rudder arm locking wire was spotted by very few even though there was a hole in the adjuster for the wire and some paint had come away when the locking wire had been taken off. These are the sort of clues you would get in reality.

All in all, this was an exercise that provoked considerable interest. We hope it will have been of help to the 200 or so people who took part with unlimited time and no pressure to get airborne and that their experience will also be of benefit to all GASIL readers.

One year later the Safety Data and Analysis Unit ran another competition at an Air Show and its summary of this event is reprinted as follows:

Preflight Inspection Competition at Cranfield PFA Rally
First, many thanks to all those who came along and had a go at finding the twelve defects. We enjoyed arranging it for you and feel that it was

a very worthwhile exercise; at one time there were over twenty people around the aircraft. We're thinking of installing a time clock next time so that it's all you can find in 10 minutes; some people enjoyed it so much that they were with us for over half an hour. As usual many of you found things we never intended and these may have put you off the intended defects, so on the Sunday we put some 'This is *not* one' labels on the aircraft. This slightly improved the Sunday results. Altogether 172 answer sheets were completed, nearly always by groups of people (7 was the biggest group), so around 500 people had a go at it, and the aeroplane survived – just!

Only three people found all 12 of our intended defects, fifteen people found 11, twenty-four found 10 and forty-four found 9. The average, with dry ground, perfect weather and good lighting, was 8.1.

When given the Answer Sheet many went to have a look at what they had missed: a useful training feature.

So what were the defects and what did people miss? (See table.)

	Fault	*Missed by*
1.	Aileron balance cable disconnected causing nasty noises and intermittent movement of other aileron and column.	26%
2.	Static vent slots on side of pitot covered by black tape.	52%
3.	Fuel filler caps labelled 'Jet A1'.	23%
4.	Left-hand door bottom hinge split pin missing.	33%
5.	*Two* screws missing from spinner.	16%
6.	Plastic bag blocking engine cooling airflow.	30%
7.	Right-hand exhaust stub removed.	63%
8.	Right-hand fuel tank contained 100% water.	36%
9.	Right-hand tyre was bald and had no creep marks.	12%
10.	Right-hand flap centre flap attachment nut was removed.	30%
11.	Rudder actuating rod turnbuckle wire locking was broken.	4%
12.	The anti-collision beacon on top of the fuselage was loose.	51%

Some points to note:
- Only two people (one of them a glider pilot) identified the exact cause of the aileron systems fault but most were highly suspicious; don't ever let anyone put you off by a statement 'they are always noisy', and remember that a full check *must* include a look at the *surfaces*. The stick was full and free in our case!
- Anyone who asked us where the static vents were on this aircraft was counted as having found the problem.
- The water in the fuel tank caused some puzzles, for there is no 'separation' when it's 100% water and the traces of fuel in the draining bottle will give it a smell. Beware of Mogas – it's clear rather like water, 100LL being blue. Some may have thought they were not expected to do a drain check, or may not have realised that the

bottles were not a 'litter hazard' but were there for a good reason. Some only drained the left-hand tank and got fuel, so didn't bother with the right-hand side.
- The flap attachment nut was hard to see when wearing sunglasses, as the shadows became completely black.

In reviewing the results of these two preflight inspection competitions it needs little imagination to appreciate the number of accidents which could have happened if the aircraft concerned had flown following these inspections ... the moral is very clear!

It is worth nothing that many in-flight emergencies have been traced back to a pilot's carelessness or lack of preparation before flight. A pilot should be acutely aware of this fact and avoid over-cursory preflight inspections.

There is little more one can say about pilot error in manipulating the various aircraft controls and systems but it would be useful to cover the insidious occurrence of carburettor icing and the way to recognise its onset, thus helping to reduce the risk of its ultimate effect – engine failure.

Fig. 10.1 Engine failures have been a hazard since the early days of powered aircraft and still are. They can occur at any time and in a wide variety of circumstances.

Engine failures have been a hazard from the early development of powered aircraft to the present day. They can occur at any time and in a wide variety of circumstances. A number of these are related to carburettor icing.

Carburettor icing accidents

On a number of occasions an engine has failed for no known reason and subsequent examination has failed to discover any fault. Under these circumstances, only two probabilities normally existed:

(1) The pilot had inadvertently mismanaged a particular control or system, e.g. the mixture control or fuel selector, and then failed to remember or otherwise give evidence of his action.
(2) The engine suffered induction (impact) icing or carburettor icing, induction icing in this context meaning airframe icing which has formed over the main induction inlet. In many cases conditions at the time precluded the formation of airframe icing and therefore carburettor icing became the number one suspect.

The following extracts from the Australian *Aviation Safety Digest* form useful reading prior to considering the technical aspects of carburettor icing. These accident reports clearly indicate the variety of operational conditions in which this type of icing sneaks in to form hazardous flight situations.

> '... The pilot of a Pawnee had been spreading fertiliser on a rice crop at a property a short distance from a town in Northern Queensland. He had commenced operations early in the morning and, using the town's airstrip as a base for reloading, had successfully completed twelve sorties to the property before stopping to refuel. Airborne again after this stop, he returned to the property once more to make two clean-up runs along one boundary of the rice field. Throughout the operation, the owner of the property had been acting as ground marker and, after the pilot had completed his second run, he flew back over the crop towards the owner to indicate that the operation was finished. The owner waved in acknowledgement and the pilot turned back towards the airstrip. But as he straightened out of the turn at a height of about 200 feet, the pilot saw that the engine was losing power. At first he was not unduly concerned as he thought he had sufficient height to reach the airstrip, but even with the throttle wide open, the engine continued to lose power. He attempted to keep the aircraft in the air by easing back on the control column but eventually the engine failed completely and he was forced to lower the nose.
> Now dangerously low, the pilot realised he would have to put the aircraft down and, seeing a gravel road below, turned steeply towards it. But as soon as he lowered flap, the airspeed decreased alarmingly

and the aircraft fell on to the road heavily in a steep nose-up attitude. Bouncing back into the air, the aircraft touched down again with the starboard wheel and wing in the long grass growing at the side of the road. Retarded by the drag of the grass and the soft, muddy soil beneath, the aircraft slowed rapidly and, when almost stopped, tipped forward on to its nose and fell on to its back. Quickly releasing his harness, the pilot kicked out the canopy side panel and extricated himself from the cockpit with only minor scratches.

Careful examination of the engine revealed no evidence of any mechanical or system malfunction and later, when the aircraft was lifted back onto its wheels, the engine started and ran normally. Although the weather at the time was fine and very warm, the humidity was high and conditions were especially favourable for carburettor icing. The symptoms accompanying the loss of power were characteristic of the formation of ice in the carburettor and, in the absence of any mechanical defect, it was concluded that ice had built up undetected in the carburettor to the point where the engine had lost all power ...'

'... Shortly before taking off on a dual training flight in a Cherokee 140, an instructor had briefed his student on practice forced landing techniques, including engine handling and the use of carburettor heat. Once airborne, the aircraft was flown to the local training area where, levelling out at 2,500 feet, the student applied partial carburettor heat and closed the throttle to simulate engine failure. Leaving the carburettor heat control in this intermediate position, the student completed his emergency cockpit drills, selected a field and established the aircraft in a forced landing pattern, warming the engine every 500 feet until the aircraft had descended to a height of 1,000 feet. Without warming the engine again, the student continued the descent until, at 300 feet, the instructor was satisfied with his performance and told him to go around. But when the student opened the throttle, the engine did not respond. The instructor immediately took control but, as the aircraft was by now very low, he had no choice but to continue with the forced landing into the selected paddock. Although the ground was wet, the aircraft touched down normally and rolled to a stop undamaged.

The cool and humid conditions existing at the time were conducive to the formation of carburettor ice and when, a short time later, the engine was started and ground run satisfactorily, it was clear that ice had formed in the carburettor during the descent and the small amount of heat selected had been insufficient to prevent it building up. When no fault could be found with the engine, the aircraft was moved to a dry part of the field and, after taking off, was flown back to the base aerodrome without further incident ...'

'... A student pilot, while on a dual training flight, was being instructed on in-flight emergency procedures. The exercise had commenced with the instructor demonstrating action to be taken in the event of fire. Starting at 4,000 feet, he had shut down the engine by closing the throttle and turning off the fuel and ignition switches. At this stage, he selected full carburettor heat. Once the aircraft was established in the descent and he was satisfied that the student was familiar with the procedure, he decided that, with the height still in hand, he would demonstrate restarting the engine by diving the aircraft.

After stopping the propeller, the instructor set the controls for a restart, and then pushed the nose down until, at 125 knots, the propeller began to windmill. But when he levelled out and opened the throttle, there was no response from the engine. Double-checking the various engine controls, he persisted with his starting attempts until, at 2,300 feet, he realised that the engine was not going to fire. Committed now to a forced landing, the instructor established the aircraft in an approach to a field only a short distance from the aerodrome and put the aircraft down without damage.

Shortly afterwards, a licensed engineer examined the aircraft but could find no defect. The engine was started without difficulty and after it had been successfully ground run, the aircraft was flown back to its base by the chief flying instructor. In the absence of any mechanical malfunction, the failure of the engine to restart was attributed to carburettor ice. With a temperature of 5°C at 4,000 feet and moderate humidity, conditions were ideal for the formation of ice and although hot air had been selected during the emergency procedures demonstration, the engine would not have been firing at this stage since the fuel and switches had already been turned off. Thus there would have been virtually no residual heat in the engine to disperse the ice which had probably formed while the engine was windmilling with the throttle closed during the recovery from the dive ...'

'... After a one-hour delay due to fog the pilot started the engine and prepared for take-off with three passengers. By then most of the fog had cleared; only small patches remained in the gullies and valleys around the strip. The temperature was 8°C with a relative humidity of about 93%. The pilot completed an engine run-up, which included a check on the operation of the carburettor hot air system, but he did not specifically check for the presence of induction ice. After completing the before-take-off checks the pilot closed the throttle and got out of the aircraft to remove condensation that had formed on the outside of the windshield. This completed, he strapped in, re-checked hatches and harnesses, then taxied a short distance to the strip and started the take-off roll.

Acceleration appeared sluggish to the pilot so he checked that the hand brake was off. The airspeed had increased to about 45 knots half-way along the strip and the aircraft just got airborne but would not accelerate any further. At this stage the pilot started a shallow turn to avoid some trees at the end of the strip and to take advantage of low ground to the left, but the aircraft settled back onto the strip. The pilot abandoned the take-off and closed the throttle, but he was unable to stop the left turn and the aircraft ran off the side of the strip. It passed through two fences and nosed over after encountering soft ground.

The pilot and passengers climbed out of the inverted wreckage uninjured but the aircraft was destroyed.

The investigation did not reveal any mechanical defect in the engine or airframe to explain the lack of performance; but atmospheric conditions at the time of the accident were certainly conducive to the formation of induction icing. With a dry bulb temperature of 8°C and a relative humidity of about 93%, the probability of serious induction ice formation would be high at any power setting. The formation of ice at low or idle power would then almost be assured ...'

Fig. 10.2

Formation of engine icing

The fact that such accidents and incidents are continuing to occur is all the more surprising when it is remembered that the problem of carburettor ice has long been recognised, the conditions under which it occurs are known, and the means of precluding it are readily available for the pilot to use at his discretion.

Carburettor ice can form under a wide variety of atmospheric conditions. It is not only a low temperature phenomenon but, as with the Pawnee in the first example, it can occur at air temperatures well above freezing point, and in clear air as well as in cloud and precipitation. Although the susceptibility to carburettor ice varies greatly amongst aircraft types, under certain meteorological conditions all piston engines are liable to be affected to some degree by icing in the induction system. In engines having a float-type carburettor, in which the fuel is introduced upstream from the throttle butterfly, icing can occur at almost any air temperature if the humidity is high enough. In the case of engines employing fuel injection systems, impact and throttle icing may still be encountered when flying in cloud or in low temperatures. The ice may form in various places and on various parts of the induction system – in the air intake, in curves of the inlet manifold, at the main jet, in the venturi, or on the throttle butterfly valve. Ice formation in the induction system can result in a loss of power, or in power not being available when needed and, in extreme cases, can render movable parts inoperative.

There are three different processes by which ice may form in engine induction systems: fuel-evaporation icing, throttle icing and impact icing.

Fuel-evaporation icing

This is the most common and insidious kind of carburettor ice and forms in the induction system at and downstream from the point at which the fuel is introduced. It is caused primarily by the rapid cooling of the induction air and the adjoining surfaces of the carburettor as the fuel vaporises.

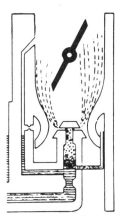

Fig. 10.3

The heat required for the evaporation process is taken from the surrounding air, lowering its temperature. If the incoming air is sufficiently humid and the temperature falls below the local dew point, excess moisture is precipitated in the form of condensation. If the temperature is reduced below 0°C, this moisture freezes. Between 0°C and about −7°C, this freezing process is relatively slow and ice will begin to form on any solid object the moisture may encounter on its way through the carburettor. In fact, at these temperatures, the moisture will even 'flow' over a surface for a short distance as it freezes, building up layer upon layer of ice as it does so. If this build-up is allowed to continue unchecked, it will eventually increase to the point where the induction passages become blocked and the engine loses all power (Fig. 10.4).

Fig. 10.4

Visible moisture in the air is not necessary for evaporation icing – only air of high humidity. It can occur in no more than scattered clouds, or even in bright sunshine with no sign of rain. Evaporation icing can normally be expected to occur within a temperature range of 5°C to 30°C, although the upper limit may extend as high as 40°C. Temperatures around 15°C should be regarded as the most suspect. The minimum relative humidity generally necessary for evaporation icing is 50%, with the possibility of icing increasing at higher humidity levels. Obviously, the icing will be more severe if there is water in liquid form in the outside air, so pilots should be particularly alert for icing during flight in rain or cloud at these temperatures. For a given relative humidity, the severity of this type of icing decreases with a reduction in temperature and it is unlikely to give trouble at temperatures below –10°C.

Throttle icing

As the induction air flows past the restriction of a partly closed throttle, its velocity increases. This is accompanied by a decrease in air pressure and a consequent fall in air temperature. As a result of this reduction in temperature, any water vapour in the induction air condenses and ice may form at or near the throttle butterfly. This kind of icing is most dangerous when the engine is running with the throttle nearly closed, such as during an approach to land or when operating on the ground. This is due to the fact that only a small amount of ice is needed to block the small gap between the butterfly and the wall of the carburettor, and there may be little or no warning that this is happening.

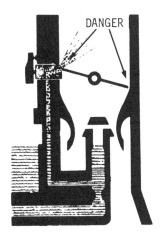

Fig. 10.5

Although opening the throttle may clear the ice, the engine could already have stopped. Remember that under these flight conditions (aircraft descending) the airflow will normally keep the propellor windmilling at around idling rpm. Even allowing that opening the throttle prior to the engine stopping may clear the ice, a heavy accumulation can make the engine slow to respond or, in extreme cases, may prevent it picking up.

The temperature drop at the throttle butterfly normally does not exceed 2°C. Thus, in a fuel injection installation, where the fuel is introduced downstream from the throttle, this type of icing is unlikely to occur at free air temperatures above about 5°C. With a normal float-type carburettor installation however, any ice formation at the throttle butterfly could be attributed to the effects of both the evaporation ice and throttle ice phenomena, and can occur at temperatures much higher than 5°C.

Impact icing

The conditions required for the formation of impact ice in the induction system are similar to those that produce airframe icing. For this reason, it is more easily anticipated than the other two forms of carburettor icing. Impact icing occurs in air temperatures between –10 and 0°C, when supercooled water droplets strike parts of the air intake or induction system which are themselves at or below 0°C. In these conditions, the water droplets, at a temperature below freezing but still in liquid form, freeze immediately they come into contact with the components of the induction system.

Supercooled moisture may be encountered in the form of freezing rain, clouds or wet snow. At temperatures above –10°C, clouds always contain water droplets and so are prolific sources of impact icing, which can build up on air intakes, heater shutters, duct walls, carburettor screens, throttle butterflies and carburettor metering elements. The most serious form of icing is clear ice which occurs readily at around −5°C, especially at high airspeeds, when there is a high liquid water content in the air, or when the water droplets are large.

Recognition and prevention of engine icing

The onset of carburettor icing is not always immediately apparent and its symptoms can easily be confused with other engine troubles. With a fixed pitch propellor, the best indication of icing in the induction system is an otherwise unaccountable drop in rpm which may or may not be accompanied by rough running. In the case of a constant-speed propellor, however, the loss of power could be serious before a reduction in rpm occurs and a more positive indication is a drop in manifold pressure or a reduction in airspeed in level flight.

Carburettor air heaters in light aircraft are usually of the exhaust pipe 'muff' type. In this system, the induction air is heated by circulating it through a muff fitted around the exhaust pipe. The exhaust-heated air is then directed through a valve into the carburettor intake or returned to the atmosphere. Under all ordinary circumstances the application of full heat will ensure that the internal temperature of the carburettor is raised above 0°C. Thus hot air can be used in a precautionary way to prevent ice forming in the carburettor, as well as to remove any ice which has already formed.

Icing in the induction system has exactly the same effect as gradually closing the throttle. The airflow through the carburettor is progressively restricted as the ice deposits build up, reducing the power being developed by the engine. If the symptoms of carburettor ice go unrecognised, pilots often try to compensate for the gradually falling rpm or manifold pressure indications by opening the throttle further and further. It is just this set of circumstances that can lead to a complete loss of power. The heat for the hot-air system is provided by the engine and if the application of carburettor heat is delayed too long, the heat available at the reduced power may be insufficient to melt a large accumulation of ice in the induction system.

It is important, therefore, that the symptoms of carburettor ice be recognised early enough for the application of hot air to be effective. Immediately icing is suspected of causing a power loss, full carburettor heat should be applied for sufficient time to restore engine power to the original level. In particularly adverse conditions, it may be necessary to apply heat for intervals of up to 30 seconds to ensure that all ice has been removed. If ice has been forming, and the use of hot air disperses it, returning the carburettor air knob to the cold position will produce an increase in rpm or manifold pressure, over the original reading.

If icing conditions are suspected during flight, it is good practice to use carburettor heat at frequent intervals to check for ice formation. If icing is severe, the control should be left in the fully hot position for as long as these conditions persist. When carburettor heat is used continuously in cruising flight, it will also be necessary to lean the engine. Hot air is less dense than cold, and the application of carburettor heat enriches the mixture. The engine should therefore be leaned in the normal way to avoid excessive fuel consumption.

A small loss of power and some rough running will always occur when carburettor heat is selected, whether or not ice has formed. This is the result of both the change from ram air to the less direct flow of the hot air system and enrichment of the mixture. If a large build-up has occurred, this normal power loss will be accompanied by an increase in engine roughness and further loss of power as the ice melts and passes into the engine as water. At first these conditions may give the impression that the application of hot air has only made the situation worse, and pilots unfamiliar with the processes involved are sometimes reluctant to apply sufficient heat for long enough to have any effect. This state of affairs is only temporary however, and it is vital that pilots resist the temptation to return the carburettor heat knob to the cold position before the hot air has had time to clear the ice. Despite the temporary roughness and moderate power loss, no engine harm can result from the continuous use of full hot air at cruise power settings of 75% or less. At higher power, however, such as during a climb, carburettor heat must be used with discretion to avoid detonation and possible engine damage. At no time should carburettor heat be used for take-off.

In aircraft equipped with a carburettor air temperature gauge, partial carburettor heat should be used as necessary to maintain safe temperatures and prevent icing. But if no intake temperature gauge is fitted, partial hot air, unless specifically recommended by the aircraft manufacturer, should never be used. The system should be either fully on or off. When operating in temperatures below the critical range for ice formation, the use of partial heat can actually cause carburettor icing by raising the intake temperature to the most critical icing range. In this situation, the temperature of the warmed air may be sufficient to melt ice particles which would otherwise harmlessly pass into the engine, but not high enough to prevent a rapid build-up of ice as the resultant moisture freezes again when fuel evaporation takes place.

Carburettors are especially prone to icing during long periods of flight at reduced power settings, such as during a let-down or an approach to land. In order to provide sufficient heat for the hot air system to function effectively the engine should be kept warm by opening the throttle at intervals of approximately 500 feet during prolonged descents. Whenever the possibility of icing exists, hot air should be selected before, rather than after, the throttle is closed, and before the engine temperature starts to fall as the aircraft descends.

Carburettor icing can also occur on the ground during taxying with small throttle openings or while the engine is idling. During the 'pre-take-off' cockpit checks, full carburettor heat should be selected, not only to verify that the hot air system is working properly, but to remove any ice which might have formed. If the aircraft is kept waiting for any length of time at the holding point, it may be necessary to repeat this procedure to keep the engine clear of ice. Except for these checks, however, hot air should not be used during ground operations. With carburettor heat selected, the incoming hot air bypasses the air intake filter and there is the possibility, especially in dusty conditions, of foreign material being drawn into the engine. Before full power is applied for take-off, the carburettor heat control must be returned to the cold position.

Virtually all accidents and incidents involving carburettor icing can be attributed to mismanagement of the engine controls. Unlike a mechanical failure, over which the pilot may have little or no control, carburettor icing can be avoided because the means of doing so are readily available. It follows that the best defence against icing difficulties is a sound knowledge of the correct method of using carburettor heat, greater awareness and vigilance on the part of pilots in learning to recognise the atmospheric conditions favourable to carburettor icing, and being alert when operating in such conditions for symptoms that indicate ice is forming.

Because of unique combinations of weather and engine characteristics, recognition of icing conditions is not always easy. Pilots should therefore be constantly alert for the possibility of carburettor icing and take the necessary corrective action before an irretrievable situation arises. It is important to remember that if an engine fails completely because of carburettor icing, it may not restart. Even if it does, the delay could well be critical.

Before leaving the subject of carburettor icing it may be of value to consider the following comments relating to a common procedure used by pilots.

During circuits and landings a number of pilots adopt the procedure of selecting carburettor heat for a few moments during their pre-landing checks. If no evidence of icing is revealed the hot air selector is returned to the cold position and the circuit continued in the normal manner. However, what is not always appreciated is that this icing check merely determines whether carburettor icing is or is not forming at the particular power setting, height and outside air temperature and humidity. Now consider the situation some 30 to 60 or so seconds later. The aircraft will be turned onto base leg, the power reduced and a descent initiated. Reducing the power will cause a significant reduction in the rate of fuel flow through the carburettor and hence a reduction in the amount of temperature drop due to evaporation of fuel. In addition, the aircraft will be descending and thus the outside temperature and humidity will be changed.

It can therefore be seen that if the temperature in the carburettor throat during the downwind leg was just below the bottom of the normally accepted icing range, i.e. 0°C to –8°C, then the action of reducing power will raise the temperature at the carburettor throat and may bring it into the icing range; this action will be conducive to the formation of carburettor icing.

The small change of temperature and humidity could also aggravate this condition. In other words the action of checking for icing on the downwind leg is no guarantee that it will not start to form during the base leg and final approach – a time when no pilot will want to find that he has an iced-up engine.

So much for the technical aspect – what now of the recognition factor in this situation? If ice is forming during the approach phase it is unlikely that any recognition of this will be possible due to the fact that the engine is under very low power (or in the case of a glide approach, idling power) and no rough running will be apparent. Further to this, the other early recognition feature, i.e. a small decrease in rpm or manifold pressure, will in most cases not be noticed. This is because the propellor will continue to windmill at around idling rpm due to the effect of the aircraft's airspeed. If a reduction in manifold pressure is the normal source of recognition for the type of engine used then it must also be appreciated that in normally aspirated engines the manifold pressure increases with decreasing altitude, and anyway how often do you check an rpm reading or a manifold pressure during a final approach?

From these comments we see that during a vital part of every flight the pilot will have very limited ability to recognise the onset of carburettor icing. The most likely time for discovery will be when the throttle is opened to correct an undershoot situation or to go around – by which time it may well be too late and an accident a certainty.

Those pilots who appreciate this fact normally adopt (dependent upon aircraft type and engines used) the selection of 'hot air' as an automatic procedure just prior to reducing power, thus guarding against the possibility of ice forming at a time when they are least equipped to recognise it. This is probably good advice for any pilot to consider, because even small amounts of carburettor ice can have an appreciable effect on the power output when a missed approach or go-around is initiated, particularly when full flap has been selected.

Figure 10.6 shows the wide range of ambient conditions conducive to the formation of carburettor icing in a typical light aircraft piston engine. Particular note should be taken of the much greater risk of serious icing under descent power. The closer the temperature and dewpoint readings, the greater the humidity. The following briefs of engine failure situations tell their own story.

> 'After losing power on the initial climb the pilot turned back towards the aerodrome. When within the boundary, the engine began to emit smoke so the pilot landed into wind on the grass. The main gear collapsed when the aircraft hit a depression in the ground. The engine had overheated and a piston had burnt out ...'

> 'Soon after take-off the engine lost power and a forced landing was carried out. The aircraft landed heavily on a steep uphill slope. Fatigue

BE PREPARED FOR CARBURETTOR ICING

Engine power losses continue to occur as a result of carburettor icing. Ice build-up in the carburettor air intake can gradually choke off the air, enriching the mixture and reducing engine power. Although more prevalent during the winter months, carburettor ice can form at any time of the year if the conditions are suitable. Learn to recognise the situation and be prepared for carburettor icing!

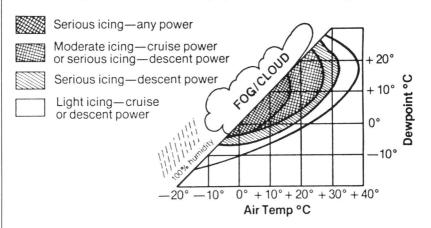

- refer to the chart when flight planning to anticipate carburettor icing
- ensure carburettor heat works during engine run-up checks (initial drop in rpm when heat applied)
- continually monitor engine instruments; loss of rpm (fixed pitch propeller) or decreasing manifold pressure (constant speed propeller) could mean carburettor ice is forming
- apply full carburettor heat early if icing is suspected, and keep it on (the engine may run rough for a short period until the ice melts)
- if the situation allows, lean the mixture carefully, after selecting carburettor heat, to smooth out the engine until the ice melts
- continue to use carburettor heat while the probability of ice formation exists, adjust mixture control accordingly
- several minutes before descent use full carburettor heat at cruise power. Periodically open the throttle during extended low power descent to ensure enough heat is maintained to melt carburettor icing and to keep the engine warm.
- if you are unsure about any of the above points discuss them with a suitably qualified person before commencing your next flight.

Fig. 10.6

failure of the brass throttle butterfly inside the carburettor throat was the cause ...'

'... On a flight from Jersey to Teeside whilst over the Channel, the oil pressure fell to zero. After a further hour's flight the engine seized and a forced landing was made. The landing gear was damaged when the aircraft ran over a ridge. After examination it was considered that the loss of oil was the result of leakage from the oil cooler connections ...'

'The aircraft's wings had been removed (in accordance with CAA AD 006–05–83) in order to renew the main spars owing to severe corrosion at the landing gear attachment.

Following the work the aircraft was ferried without problems to an aerodrome about 100 miles away, where the owner collected it. About 50 litres of fuel were added to each of the four fuel tanks (which had been removed during the wing spar renewal). The gauges showed rather more than half full (capacity 100 litres per tank), but were known to be inaccurate in the middle of their range. During the one-hour flight the pilot noticed that three out of the four fuel gauges were showing no small movements, leading the pilot to suspect they were sticking. While descending through 1,000 ft about one mile from his destination, the engine stopped. He was able to restart it by selecting the tank with the fuel gauge which was acting normally. Three of the four tanks were found to be collapsed, jamming the fuel gauge floats. The tank vent pipes are 3 mm internal diameter plastic tubes which originate on the inside neck of the fuel fillers and emerge under the wings, where they project downwards about 2". The ends of the tubes are cut off at an angle of 45° and three of these had been incorrectly cut with the angle facing the rear of the aircraft. The suction was sufficient to cause the tanks to be emptied, which could not be seen because the low wing, before collapsing. It is not known how much fuel was on board during the first ferry flight but that pilot was probably lucky to have selected the tank with a correctly finished vent pipe ...'

'On a solo flight following a dual check, the pilot took off on runway 07 for a circuit detail. The wind was calm and the temperature +5°C, with no visible moisture.

The pilot had been briefed to carry out three touch and go landings followed by a final full stop landing. One landing and take-off were successfully completed but, as the aircraft was rotated at 60 kt for the second take-off, the engine lost 300 to 400 rpm and the aircraft sank back onto the runway. The pilot closed the throttle but, despite the maximum possible braking, the aircraft overran the end of the runway and collided with the perimeter fence. There was no fire and the pilot escaped without injury.

Subsequent examination and test running of the engine failed to identify any fault or reproduce the power reduction ...'

'Following one touch and go landing, the aircraft made a normal approach to a full stop landing. On initial touchdown at the runway threshold the aircraft bounced and the pilot applied throttle in order to cushion the second impact. The engine failed to respond and the aircraft made three or four successively heavier landings before coming to rest.

On three previous occasions the engine had cut after the landing and, following inspection, it was considered that carburettor icing had been the cause. It has been subsequently stated by the pilot that during this flight there was no chance of carburettor icing.

The weather was reported as CAVOK with a temperature of 15°C...'

Even bad landings can contribute to engine failure...

'The aircraft was being flown by a student pilot for his qualifying cross-country certificate, from Lashenden to Goodwood. The landing at Goodwood was very heavy and the aircraft bounced. The pilot initiated a go-around and observers on the ground at Goodwood reported that the noseleg appeared to be damaged. They suggested that the pilot return to his home base at Shoreham. This was accepted by him and during the flight back to Shoreham, the pilot of an accompanying aircraft advised that fuel could be seen leaking from the area of the damaged nose undercarriage mountings.

The pilot arrived at Shoreham but on the downwind leg for Runway 31 the engine lost power, apparently due to fuel starvation, and he elected to land on Runway 25 instead. The aircraft touched down at the extreme end of Runway 25, the noseleg collapsed and the aircraft came to rest in the overrun area. The pilot evacuated the aircraft without injury.

Subsequent inspection indicated that the heavy landing at Goodwood had bent the nose wheel back and distorted the firewall sufficiently to partially fracture the pipe feeding the 'Gascolator' (fuel filter) mounted thereon. The fuel tanks were found to be empty of usable fuel...'

'In normal cruising flight at 1,500 ft altitude, in clear weather with a cloud base of 3,000 ft and a temperature of 9°C, the aircraft's engine began to suffer rough running. The pilot headed the aircraft towards the departure airfield, Rochester, but the engine lost more power and a forced landing was carried out in a field. Wind shear was experienced on the final approach and the landing was heavy, causing damage to the undercarriage and propeller.

Carburettor hot air had been applied initially but when the engine roughness increased it was de-selected. An examination of the engine and its fuel and ignition systems had not revealed a defect which could explain the power loss...'

(*Comment.* Though the reason for the engine failure here is not known, when carburettor icing is experienced initial application of carburettor hot air often does result in increased rough running.)

'The aircraft was engaged on a local flight from Goodwood aerodrome when the engine stopped. During the forced landing at

Hayling Island the aircraft was damaged beyond repair and the pilot seriously injured ...'

'The aircraft suffered a double engine failure on joining the circuit at Biggin Hill. It force landed in a field adjacent to the aerodrome and was subsequently damaged. The occupants received minor injuries. The aircraft was out of fuel ...'

'The aircraft was just completing a flight from the Isles of Scilly to a landing strip near St Ives when, after letting down from 1,500 ft and when on base leg, the engine failed. The pilot immediately changed heading direct to the landing strip, but found it necessary to raise the aircraft's nose to clear some trees on the south-west corner of the airfield. As a result of speed loss from this manoeuvre the aircraft stalled when about 25 ft above the runway. It struck the ground tail first before pitching over onto its back. There was no fire. The pilot reported that at the time of the accident 150 lb of fuel were on board the aircraft.

Subsequent investigation by the pilot into the cause of the engine failure revealed a significant quantity of water in the fuel lines to the engine, and evidence of water in the left fuel tank.

Prior to this flight the pilot reported the aircraft had been parked, with both tanks approximately half full, on a slope with the right wing lowermost. During this time an imbalance in the tank fuel quantities has arisen by, he considered, fuel transference through the tanks vent system.

Due to weight considerations the pilot elected not to refill the depleted tank, knowing that there would be sufficient fuel on board for the intended flight. In retrospect, he considers that there was insufficient fuel in the depleted tank to carry out a satisfactory preflight water drain check ...'

Reducing the risk of engine icing

Due to the variable factors involved in the formation of carburettor ice, e.g. outside air temperature, adiabatic cooling, evaporation cooling and atmospheric humidity, it is not easy for the pilot to forecast when icing will occur. He must therefore remain on guard against such a possibility occurring during every flight. The remedial actions which are available whenever icing does occur are as follows:

- Use full heat to clear any icing which has formed, then, as recommended by the aircraft manufacturers, use partial or full heat to prevent recurrence. If partial heat is used the amount of heat will have to be determined by trial and error unless a carburettor air temperature gauge is fitted.

    Alter the power setting or if applicable use mixture control. The evaporation of fuel normally accounts for some 70% of the temperature drop in the carburettor and therefore a power change can often have a

positive effect upon the internal temperature of the carburettor. The greatest effect upon temperature change is usually made by increasing power.
- Alter the altitude at which the aircraft is flying. This, however, is usually a somewhat restricted remedy against ice reforming as the normal outside air temperature change is only 2°C per 1,000 feet, and the constraints of a 'safe height above the surface' when reducing altitude and the fairly widespread existence of 'controlled airspace' when increasing altitude do not always allow for much altitude variation.
- The carburettor heat control should not normally be used during taxying unless ice formation is suspected, since the air from the carburettor heat system is normally unfiltered. Taxying with unfiltered air could cause dirt, grass cuttings, etc., to get into the engine cylinders and cause the cylinder walls and piston rings to develop excessive wear.
- When applying carburettor heat as a precaution, partial application may tend to induce carburettor icing by warming air which was below the temperature range for ice formation and raising it up into the ice formation range. The Aircraft Manual must be checked to determine the correct procedure for the aircraft type.
- Remember that in the event of airframe icing which forms over the induction inlet, the use of carburettor heat will automatically provide an alternative air source, thus preventing any significant power loss or an engine failing from this type of icing condition.
- Most owners' handbooks and aircraft manuals warn against prolonged engine operation on the ground with alternate or hot air selected, as the air source is unfiltered. However, during run-up (and in icing conditions immediately before take-off) hot air must be applied for sufficient time to ensure that any ice that may have formed is removed. The technique will vary with different aircraft types so check your owner's handbook or aircraft manual for the correct procedure for your aircraft. Generally, application of hot air will cause an rpm drop. If no ice is present the rpm will remain steady at the lower figure. If ice is present the rpm will initially decrease, then increase as the ice is removed, and stabilise at a higher reading. The rpm will then increase further when cold air is again selected. This is a worthwhile check following any prolonged period of operation on the ground in possible icing conditions. Make it your habit to check for the presence of ice as well as checking the serviceability of the carburettor hot air system.
- When full hot air is being used the pilot must bear in mind that the engine power output is reduced and the fuel consumption rate is significantly increased and could reduce the aircraft range by as much as 20% without proper adjustment of the mixture control.

The forced landing situation

Having covered some common causes of engine failure it may also be helpful

Fig. 10.7

to touch upon some points of procedure should such an event happen to you. In single-engine aircraft a powerplant failure will, to most pilots, be a highly traumatic experience. If the failure is sudden and unexpected the psychological shock of knowing that the only direction to go is downwards will easily induce a mental block to clear thoughts and organised actions.

How will you respond?

For a moment, just consider a typical situation as it happens:

> The weather may be good, the aircraft may be climbing, cruising level, or just being levelled off following a descent – the pilot is occupied with his thoughts, his navigation, or talking to his passengers, when suddenly into the cockpit comes a deathly hush – it has happened – the faithful engine up front upon which he has come to rely has gone on strike. To a pilot sitting at 3,000 feet this is the quietest moment in the world!

It is now that clear thinking is at a premium; chaotic thoughts produced by panic are of no value, a level head and quick planning offer the only way out. It is in this situation that the pilot assumes the fullest responsibility of captaincy: no longer is he a navigator, or a driver, it is here that the decisions must be *right* and he must make the most of every opportunity to get his passengers and himself down on the ground safely.

> The aircraft sinks, the altimeter begins to unwind and at the back of his mind the pilot realises that this is the situation which his instructor had

been preparing him for during those practice forced landings without power, but this time it's not practice, it's for real!

Even with clear thinking and good planning the results may be unsuccessful, and there is certainly no time to make errors or allow muddled thinking to interfere.

Have you ever considered the psychological effect of sitting at 2/3,000 feet grasping a throttle which is connected to an unresponsive engine? Well, what indeed would be the effects on you? Philosophical thinking, panic, seizure of the mind, beads of sweat on your forehead – the stage is set for any or all of these.

Reflecting upon the above words, did you fully realise, during your training, the vital importance of practising for such a situation? Did you fully appreciate the problems that such an event would bring about? Did you analyse the degree of stress it would place upon you? And have you maintained, through regular practice, a reasonable degree of competence to cope with such an occurrence? Or are you a pilot who just accepted 'forced landing practice' as a training exercise which had to be covered in order to complete your course and obtain your pilot's licence?

Having said all this, it must be appreciated that, during basic training, it will take a fair amount of flying practice to become really proficient at handling a forced landing situation, and it is difficult to find this time in the normally accepted period for completing a private pilot course. Further to this there is also the fact that even though competence has been achieved and maintained it may not prevent a hazardous arrival on the surface if the available area is a hostile one, e.g. mountainous terrain, water, built-up area, etc. It is for this latter reason alone that it is so vital to avoid human failure in the handling of the engine and its systems.

Planning for success

With reference to the best procedure to adopt in these circumstances, it should at least be sufficiently simple to be carried out under stress, and at the same time afford a good opportunity of success. In this respect it can be said that for any two pilots of equal ability, experience and judgement, the one with the greater confidence in himself at the time will be the only most likely to succeed.

It can be seen from this statement that confidence in oneself has a very large effect on the degree of success achieved and as such it plays a very important part throughout the recurrent practice which all pilots should undertake on suitable occasions. This confidence can be acquired in different ways but two good examples are given in the following paragraphs:

(1) It can be engendered by the frequent practice of closing the throttle for a few seconds during various phases of flight and at a safe altitude. During these few seconds you should practise the immediate adoption

of a gliding attitude, retrimming the aircraft and assessing the wind direction, whilst searching for a suitable area for a forced landing.

This activity can be practised fairly often without inconvenience to the nature of the particular flight. These actions are more important than many people realise, because it is in the first few moments following an engine failure that seeds of panic are most likely to burst into their embryo form and result in loss of time vital to planning, and if this danger is reduced we have gone a long way to the eventual success of a safe landing.

(2) Another 'confidence builder' is the selection of a suitable landing area. Consider for a moment that when an engine fails the area below you consists of miles of flat pasture land extending in all directions. In these benevolent circumstances, your confidence will be high because all that is required is for you to determine the approximate wind direction, take up an appropriate heading and glide down to a normal landing into wind.

Now, to encounter such circumstances would be rather uncommon but there will be occasions when an aircraft which has suffered engine failure is above terrain where relatively large open areas exist, broken only by hedgerows. In this situation the pilot should choose a suitable field somewhere in the centre of this area, thus knowing from the start that if he misjudges the aircraft positioning during the descent he will still have the opportunity of re-selecting a landing path into a different field. This facility would also be a vital one should he find, on nearing the ground, that the original field was unsuitable because of small streams, ditches, etc., which were not visible from the height at which the original selection was made.

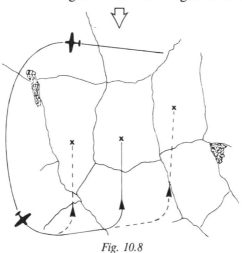

Fig. 10.8

The accent here is on flexibility, because the choice of one field alone without consideration of the immediate surrounding area lacks correct planning, in that no alternative may be available if the first choice is found to be unsuitable or miscalculations occur on the way down – and as we know, the latter is a very real possibility.

However, whilst 'area selection' provides escape routes it should be used with care and any decision to change the landing point should be taken whilst sufficient height remains to permit a change in plan.

Finally, on the question of area selection, pilots should occasionally force a change of plan on themselves during practice in order that they can develop at least some ability to re-select a landing path should circumstances necessitate. Above all we must remember that practice in forced landings is not an academic exercise, it is to prepare oneself for the real thing, should it happen!

In order to plan the descent pattern and final approach successfully, the approximate wind direction must be known and in this connection there are only three practical methods open to a pilot:

(a) Indications given by smoke.
(b) The known take-off direction, when flying in the local area.
(c) The known forecast wind, when cross-country flying.

In any event the approximate wind direction relative to the aircraft heading should always be kept ready and available in a pilot's mind. Some reference books refer to cloud shadows, ripples on water or crops, and cattle grazing in the lea of a hedge, but in this emergency situation you will almost certainly not have time to study the ground for cloud shadows – and if you do find them, will you really be in a state of mind to work out the correction to be applied in order to find the surface wind? Again, to attempt to find the wind direction from fields of waving corn is more poetical than practical, and the eating habits of cattle have little relevance to wind direction.

In relation to the 'descent plan' there are various methods taught by instructors and developed by pilots. However, regardless of the actual technique preferred, e.g. the spiral from overhead, or the rectangular pattern around the landing site, the two most vital factors will be the need to maintain a safe airspeed and the requirement to continuously monitor the aircraft's position relative to its height above the surface, and its distance from the intended touchdown point, measured along the descent path.

In this respect and regardless of the descent method used a simple aid to judgement in positioning the aircraft for its final approach is to bear in mind that you will normally need about 1,500 feet above the ground as you commence that portion of the descent which constitutes the short downwind leg section, whether it be along a straight or curved path (see position 4 in Fig. 10.9).

Engine failure – remedial actions

Once the descent has been initiated, the normal practice is to attempt to find

PREVENTION IS BETTER THAN CURE

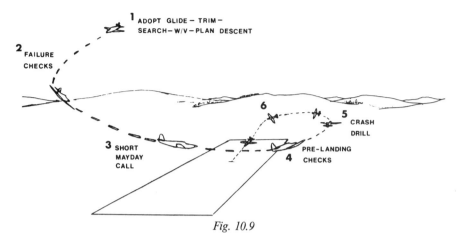

Fig. 10.9

the cause of engine failure and if possible rectify it. During training and subsequent practices a pilot's mind is prepared and alert to these checks, and thus they are not too difficult to remember in such situations. However, in the case of an actual engine failure the atmosphere in the cockpit is very different and the events which follow can easily overwhelm the pilot's thought processes. This can lead to some of the checks being overlooked, or carried out too cursorily for the cause to be determined and subsequent rectification achieved. This fact has been demonstrated in a number of past accident reports.

Pilots must therefore be cognisant of this fact and take care over these important checks. They should be performed one at a time, and in between monitoring the aircraft's flight path relative to the intended landing path. Regardless of the order in which they are done, the more usual checks will consist of: fuel, carburettor heat or alternate air systems, mixture, ignition and the oil pressure and temperature gauges.

The causes of engine failure which may be rectified are not always easy to demonstrate in a realistic manner during practice and this should be borne in mind. For example, taking just two past occurrences involving engine failure caused by inadvertent interference with the fuel selectors:

(1) In this case the woman passenger sitting in the right-hand seat of an aircraft type which is used extensively in general aviation placed her handbag on the floor by her left foot. Later on during the flight she reached down and picked up her bag, which had a shoulder strap attached to it. In doing so the handbag strap caught underneath the vertically-mounted fuel selector and turned it to the OFF position. At the time of this occurrence the aircraft was in a descent towards the destination airfield and the engine was throttled back. The pilot only discovered the engine had failed when he opened the throttle to level off at circuit height. During the very hasty failure checks the pilot omitted to verify the position of the fuel selector lever as he had only just double-

checked this during his aerodrome approach checks...

(2) A second example concerned a twin-engine aircraft in which a rather long-legged passenger was sitting centrally in the seat immediately behind the front seats. At some stage the passenger stretched out his legs and the heels of his feet moved forward between the two front seats. In this aircraft the fuel selectors were sited on the floor between the two front seats. It takes little imagination to understand what occurred next. The passenger's heels moved both fuel selectors forward into the OFF position. Thus both engines failed within a few seconds of one another and at a time when the pilot was in the process of readjusting the mixture control levers to a more lean position as the aircraft had just levelled off from a climb. The pilot automatically assumed the failure of both engines was related to his moving the mixture controls and during the subsequent checks failed to verify the position of the fuel selectors...

In relation to engine failure as a result of misuse of the mixture control a likely cause is that the pilot has been flying high, with the mixture adjusted for height and the knob well out. If a descent is now commenced and the mixture knob position forgotten there will be trouble when the aircraft is levelled off at a substantially lower height. Remember that no rough running or lean mixture symptoms are likely when gliding or descending at reduced power, and in the worst case, application of high power when levelling off could result in the engine cutting out completely.

Again, a similar situation could arise with carburettor icing. If icing occurred during a power off, or reduced power descent, the two main symptoms, i.e. loss of power and/or rough running, are unlikely to show and it is only when the throttle is opened again to cruising power at the end of the descent that the engine will not respond.

Turning to the ignition area we should realise that whereas it is normally very unlikely for switches to be knocked off, the key-type ignition system does have at least two features which can lead to the running of an engine on one magneto only. The first is the feature on most aircraft where the words OFF – RIGHT – LEFT – BOTH are printed on the instrument facia which is separate from the ignition key barrel; if rotary movement of the ignition key barrel takes place at any stage then the ignition key selection could well finish up on a single magneto instead of BOTH. It may be said that if a magneto check had taken place before take-off this would not happen – however, it could have been this very check which caused the ignition key barrel to move.

Again, there have been many instances of pilots who on completion of the engine run-up have returned the key to the LEFT magneto instead of BOTH, which would again mean flight taking place with only one magneto switched on; failure of this magneto would cause engine failure which could be remedied by simply switching to BOTH.

Before leaving the various reasons which could cause an engine to fail, it would be appropriate to consider the question of fuel flow in relation to high power settings, such as when taking off or climbing. Aircraft fuel systems are designed so that the bore of the fuel pipes is sufficient to maintain the maximum rate of fuel demand at maximum power. However, should the fuel lines suffer a partial split or a blockage in the filters between the tank and the carburettor, the amount of fuel flowing to the engine will decrease. This could lead to a situation where the engine will stop if the throttle setting is too great for the amount of fuel flowing through the lines, but might continue running if a smaller throttle setting is used. From this it can be seen that in the event of an engine failure occurring for no apparent reason during operations at high power, it might be of advantage to select a significantly lower power just in case the engine can be kept operating at a sufficient power to maintain the aircraft in level flight, or in any event be of some value by giving partial assistance during a forced landing.

A word of warning: in the event that an engine can be kept running at a power which is insufficient for level flight to be maintained, it will be inadvisable to rely upon its assistance during the final stages of the forced landing procedure. This is because the source of the problem may be worsening and the engine could fail completely at a critical stage of the approach.

With reference to the final stage of a forced landing, it is significant that approximately 20% of the stall/spin accidents occur during these circumstances. This is the result of misjudgement leading to an undershoot condition on final approach. In a number of cases the undershoot could be traced back to the use of too much flap, too soon. Further to this it should be noted that flap selection should, where possible, be avoided during the final turn onto the approach. Use of flap at this stage will often result in a substantial speed loss during the turn and a dangerous situation at low height could result.

The final 'full flap' selection should be left until assured of reaching the original aiming point, assessment of which will have been made at height, and which should have been about one-third of the distance into the field measured from the downwind boundary. This final flap selection should be used to drag the original touchdown point back to one-third or one-quarter the distance into the field (dependent upon the landing distance available) and thus allow the longest safe landing run.

The important factor to cover in this respect is the rather indefinite appreciation of wind gradient. It should be stressed that regardless of the experience of the pilot concerned it is very difficult to judge the wind gradient effect during the last few hundred feed, and this difficulty is automatically increases with stronger winds and a strange (non airfield) terrain. The wind gradient is particularly difficult to assess when conducting a landing away from a relatively flat area like an airfield. This must also be appreciated by all pilots during practice forced landings, particularly when climbing away following the approach. Light aircraft can continue sinking under full power when winds of 20–25 kts are experienced and wind shear or turbulence is present, in the region of woods or over undulating terrain.

When practising simulated forced landings, remember one extremely important point: you are practising to keep your judgement toned up should an engine failure actually occur. Therefore when selecting a 'simulated landing area' do make sure that you have a relatively long overshoot area, one well clear of obstructions, e.g. woods, power lines, etc. The reason for this is contained in the reason for your practice: an engine may fail at any time and for different reasons, and thus it may be that just when you are climbing away after your practice the engine decides to turn that practice into a real forced landing! The statistics tell us that 20% of all accidents occur because of engine failure or malfunction. There is little sense in joining the statistics just when you are preparing yourself *not* to become one.

Fig. 10.10

Reading through the few accident summaries which follow will clearly show the need for pilots to find time at reasonable intervals to practise forced landing planning and procedures from various altitudes and in different circumstances.

'Whilst flying at 1,200 ft en route to Barton Airfield, Manchester, the engine lost power, which was accompanied by a violent vibration. The pilot selected a field and started an approach to land when he noticed a building partially obstructing his landing area. The pilot realigned his approach path to land in a field next to the one he had originally chosen. The touchdown was normal but at the end of the landing roll there was a steep downslope to a dry stone wall and a minor road. The left wing tip and undercarriage struck the wall, which resulted in the aircraft pivoting anticlockwise, tipping onto its nose and breaking the undercarriage. Subsequent examination of the engine revealed that the crankshaft had failed ...'

'The pilot, who was also the owner and constructor of this aircraft, took off with the intention of conducting a local flight. After climbing to approximately 700 ft, and when positioned at the start of the downwind leg of the circuit for R/W 14, the aircraft's rate of climb reduced markedly and the engine cylinder head temperature (CHT) rose. A short while later, as the engine noticeably lost power and the CHT continued to rise, the pilot decided on an immediate landing and headed for the threshold of R/W 14. However, it soon became apparent to him that the aircraft would not be able to glide that far so he elected to make a forced landing, into wind, in an adjacent cornfield. On contact with the standing crop the aircraft pitched over onto its back, but the pilot escaped without injury and there was no fire.

The engine fitted to this aircraft was basically a converted 1834 cc Volkswagen car engine. The pilot considers that the cause of the failure may be attributed to the air/fuel mixture ratio being too lean, although after making adjustments to the carburettors prior to this flight the engine had been ground run for several hours, during which time it performed normally ...'

'A student pilot was engaged on a circuit detail when the engine failed at 450 ft. A forced landing was made in the Barry Island holiday camp car park but the aircraft ran through railings at the perimeter of the car park and dropped about 100 ft onto the beach below. A passer-by released the seriously injured pilot from the wreckage. The full-harness probably saved his life ...'

'The pilot was an experienced helicopter pilot who was undertaking a training course on an autogyro. He had been authorised for a navigation exercise, and after take-off from Shipham airfield had turned left before leaving the circuit. There had been a crosswind during his take-off and this left turn resulted in his now flying downwind. He experienced at this time an apparent power reduction and he tried selecting the carburettor air intake to 'Hot' without gaining any improvement. Full power was selected but, with a reduced airspeed, he was only just able to maintain height and he decided to carry out a precautionary landing in what appeared to be open pasture. It was, in fact, a crop of early wheat and after touching down on the mainwheels at a low forward airspeed the nose wheel touched and then sank into the soft soil causing the aircraft to pitch forward until the main rotor struck the ground. The aircraft rolled onto its side but the pilot was able to step out without injury ...'

'At the end of a short local area flight, the aircraft was returning to the field for a landing when the pilot noticed a drop in engine rpm, which he attributed to throttle creep. However, as the aircraft approached the landing area, the engine stopped and the pilot made a forced landing in a ploughed field.

As the surface air temperature was 0°C, the pilot now attributes both the drop in rpm, and the subsequent engine failure, to carburettor icing ...'

'The student pilot had carried out two practice forced landing exercises without incident and he was descending through 2,000 ft on his third, when he smelled strong acrid fumes which he thought were coming from the engine. He opened the DV panel to ventilate the aircraft and put on full power to climb away. The fumes persisted, so he closed the throttle and began an emergency descent. He elected to make a forced landing as he feared that he would be overcome by the fumes. His touchdown speed was high and he bounced before settling the nose and port main wheels. The aircraft overran the selected field, crashed through a hedge and came to rest in a minor road after having turned through 180°.

A technical investigation revealed no defect that could have caused the fumes described by the pilot. However, there is a brickworks 6 nm upwind of the crash area and the pilot may well have smelled fumes emanating from the works ...'

'The aircraft departed from Schipol for Biggin Hill in good weather with the pilot and two passengers aboard. It climbed to Flight Level 80 on an IFR plan, where it cruised clear of cloud and in good visibility. The flight was uneventful and the pilot was cleared to descend about 30 miles before Detling VOR. It crossed Detling at 2,500 ft and set course for Biggin Hill, maintaining 2,500 ft. When about 4 miles from Biggin Hill, the left engine started to surge. The pilot switched on the electrical fuel pump and the engine picked up but failed again after about a minute. He feathered the left propeller and noticed that both fuel gauges were reading just above zero. As he continued towards the field the right engine rpm started to fluctuate but, unlike the left engine, selection of the electrical pump did not restore power.

The pilot realised that a forced landing was inevitable and so he selected a large field that was orientated into wind and attempted to land in it. Unfortunately, the aircraft was too low and it touched down 30 yd before the boundary hedge of the selected field. The aircraft came to rest with the nose and engines embedded in the hedge. There was no fire.

Neither of the front seat occupants had worn the upper torso restraint harness that was provided. However, the pilot, who was only slightly injured, managed to escape through the front door.

Examination of the aircraft revealed no pre-existing defects that could have caused the accident. The fuel tanks were undamaged and a total of $\frac{3}{4}$ of an imperial gallon of fuel was drained from the fuel system.

The pilot stated that he thought he had sufficient fuel for the flight and he did not consider that he had a fuel problem until the left engine rpm started to fluctuate ...'

'On his preflight inspection the pilot estimated, by dipping the tank, that there was adequate fuel for 45 minutes' flight in the left tank and for 30 minutes' flight in the right tank. After taxying on the right tank, the pilot switched to the left tank to take off for a flight to Nottingham airport (estimated flight time 10 minutes).

At 1,500 ft in level flight the engine stopped without any prior misfiring or uneven running. A forced landing was carried out in a ploughed field. The nose wheel collapsed and the aircraft tipped over onto its back ...'

'At approximately 40 ft agl an instantaneous power loss was experienced, the pilot immediately lowered the nose to increase speed and the aircraft descended rapidly. The aircraft did not stall, and a successful flare and landing was accomplished, but the ground speed was too high to permit the pilot to stop or turn away from the boundary hedge (the aircraft was not fitted with brakes). The aircraft entered the hedge and struck a tree stump, damaging the airframe ...'

11
Stall/spin accidents

No book written on the subject of reducing risk whilst flying would be complete without making some comments upon the accidents caused by a stall or a spin. It is also necessary to bear in mind that the aviation authorities of different countries and also those people responsible for pilot training have diverse philosophies of how to train pilots against entries to the inadvertent stall/spin situation.

Regardless of the different viewpoints of instructors and others there is one simple fact that cannot be argued with, and that is, 'if a stall is prevented, a spin will not occur'. Thus, in considering how one can guard against the inadvertent spin and stay safe, it is clear that the real objective is to avoid the inadvertent stall.

To get this into perspective it is necessary to appreciate that all stall training has one final objective: to train a pilot to avoid ever entering an inadvertent stall. To be able to recover from a stall with minimum height loss is a worthy aim at the commencement of stall training, but by itself it is not enough, as has been seen in the past from the numbers of actual stall/spin accidents which continue to occur.

The past record

The Aircraft Owners and Pilots Association of the United Kingdom commissioned a Working Group to evaluate the effectiveness of stall/spin training in the UK and the results of this study clearly highlight the positive need for a pilot to be trained in 'awareness and avoidance' techniques rather than just being taught to recognise stalls and spins when they occur and then recover from them. The Study Group carried out a detailed analysis of the stall/spin accidents which had occurred in the UK over a 20-year period and an extract from its report concerning spinning accidents is as follows:

> '... Between 1960 and 1979 inclusive there were 61 spinning accidents. Of these, 6 were the result of intentional spinning and 55 were unintentional spins. The total spinning accidents were then split into 2 pilot experience groups, one group related to pilots with less than 800 hours and the other to pilots with 800 hours or more.

STALL/SPIN ACCIDENTS

The 20-year record, unintentional spin entries

Height (feet)	No.
Ground level up to 100	3
100 to 200	5
200 to 300	7
300 to 400	6
400 to 500	5
500 to 600	5
600 to 700	4
700 to 1000	11
Total	46

Height/altitude unknown at entry Causal factor	No.
Weather	2
Air display	4
Glider cable	2
Obstruction to controls	1
Formation flying	1
Air test	1
Photographic flying	1
Forced landing	3
Total	15

Grand total 61
7 with an instructor on board
A total of 55 occurred between the surface and 1,000 ft agl.

The Study Group noted that 46 out of a total of 61 spinning accidents occurred at heights insufficient for recovery. Additionally, from the first table we see that 4 occurred during display flying, 2 involved a faulty glider cable release and 3 occurred following engine failure when the pilot was in the process of making a forced landing. It can fairly accurately be assumed that all these entered spins at a relatively low height, which then brings the figure of 46 up to 55. Thus is appears that 90% of inadvertent spin entries occur at heights where the pilot would not have the opportunity to recover even if he were competent to do so.

The vital message which can be gleaned from this fact is that, regardless of how well pilots are trained at recovering from developed spins, the majority of UK spinning accidents occurred because the aircraft entered a spin at a height which was too low for a recovery to be effective. This indicates an additional need for training methods to be aimed at developing the ability to recognise impending spin situations and effecting a recovery at the incipient stage.

Total spinning accidents

Pilots with under 800 hours' experience

Causal factor	No.
Weather	6
Engine failure	3
Low level aerobatics	4
Wake turbulence, obstruction to controls, and faulty glider release	5
Various other causes between 200' and 700'	18
Total	36

Comment: Regardless of how well these pilots had been trained at recovering from a developed spin, none had much chance (if any) of recovering safely.

Pilots with 800 hours or more

Causal factor	No.
Aerobatic display	4
Midair collision	2
Medical	1
Other causes	8
Too low to recover	10
Total	25

Comment: 4 out of the 25 occurred at an altitude sufficient to effect a safe recovery, the remaining pilots did not stand a chance.

Spin recovery – training limitations

This really means that because the normal entry to a spin stems from a stall, pilots should be better trained to recognise incipient stall situations and return the aircraft to a safe condition before the stall occurs. This philosophy is strengthened by the fact that most modern aeroplanes are not cleared for spinning when in the Normal Category. This means that during training the CG is further forward than when carrying passengers. As a result, and due to the clean lines and steep autorotative attitudes of the modern aeroplane, care has to be taken not to move the control column too far forward during the spin recovery action. In practice this forward movement of the control column will often be very small due to the forward position of the CG when in the Utility Category. A large forward movement will normally cause an extremely rapid increase in airspeed due to the steep attitude of the modern aircraft during a spin, and increase the possibility of overstressing the aircraft during recovery.

The student pilot is therefore 'habit trained' not to move the control column very much during his recoveries from developed spins (if indeed he is in a proper developed spin in the first place). If on the other hand a developed spin occurs when the aircraft is in the Normal

Category, i.e. at a greater weight and further aft CG, a subtle but important difference to the spin recovery will take place, i.e. an extremely firm and positive push forward on the control column will be required – something for which the pilot has not been habit trained. Therefore the motor skill which has been developed during training is unlikely to produce the effect sought, unless the aircraft is in the Utility Category at the time of the unintentional spin.

As only 11% of the Flying Training Organisations in the UK operate aircraft which will require a positive and firm forward movement of the control column to recover from a spin when used in the training mode, most student pilots are unable to receive the 'habit training' in relation to control forces required for recovery from a properly developed spin and this could be an important adverse factor in the results column of the training they are given.

Thus, at the risk of repetition, we are left with the message that there is a positive need for pilots to improve their recognition of situations in which the airspeed is low and close to a stall condition. The problem with all forms of stall recognition training is that for safety reasons practice in this exercise has to take place at a height in the region of 2,000 to 3,000 feet, or higher above the surface. Yet we have seen from the AOPA Study Group report that the majority of stall/spin accidents occurred at very low heights. Most of us can appreciate that it is not a difficult matter to recover from practice stall situations when we are prepared for them and they are conducted at relatively safe altitudes. But such an environment is quite different to stalling at low height and in a state of unpreparedness. It is this latter situation which causes a double exposure to hazard, in that the pilot is frequently caught off his guard by the unexpectedness of the stall and secondly there will be a strong intuitive resistance to moving the control column forward, due to the proximity of the ground.

Before further comments are made concerning this type of situation it would be beneficial to recall the proper method of stall recovery, which in its entirety is as follows:

Move the control column forward, holding the ailerons neutral whilst smoothly applying full power and preventing yaw with rudder.

To a pilot in the final stages of an approach who has unwittingly got into a stall, the recovery procedure as outlined above is akin to suddenly asking you to drop this book, jump to your feet, hop up and down on your right foot whilst patting your head with your left hand and rubbing your stomach in an anticlockwise direction with your right hand – yes, you will get there in the right sequence in the end – but to a pilot in a stall below 500 feet, the end is but a few seconds away!

You may say that statements like this are not much help in reducing the risk and up to a point you are right, but only if the matter is left like that. The comments in the previous paragraph are made to direct attention to the problem and, clearly, if we wish to have a better grasp of how to reduce the risk, we must first assess the difficulties involved in this particular situation and then by examining the factors which influence the risk, see what, if possible, can be done.

What is immediately obvious is the difficulty in coping with an unexpected event of this nature so close to the ground. It is further evident that some form of distraction must have occurred for the pilot to get into this situation, i.e. failing to monitor the airspeed at a vital stage of the flight. The subject of distractions has already been mentioned earlier in this book and these comments relating to stall/spin accidents only serve to emphasise the need to stay ahead of the aeroplane at all times. In this case it means being sufficiently competent to recognise at an early stage that a low speed situation is developing. This recognition should be accomplished not only by reference to the air speed indicator and stall warning devices, but also through the pilot's sensory perceptions. It is for this reason that the practice of 'slow flight' forms an important part of the flight training curricula of many countries.

We will now analyse a situation where, due to distraction or other cause, a stall has inadvertently occurred, and it may be of help to be forewarned of the psychological and physical problems which a pilot will meet with when it happens close to the ground. If we return to the standard method of recovering from a stall and consider the various physical actions that we are taught to use in order to effect a recovery, it will be clear that all these actions have their own specific value, but one of them is absolutely vital to an effective recovery, and that is the forward movement of the control column, which reduces the high angle of attack which led to the stall in the first place.

Taking the other actions in turn, their effects can be described as follows:

Power

This will certainly be needed if height loss is to be kept to a minimum. It will also make the elevators and rudder more effective, an aspect of particular importance when height is at a premium. It could, however, have two positive disadvantages if not properly synchronised with the action of moving the control column forward. For example, the application of power on its own will cause the nose to rise and increase the angle of attack, thus aggravating the stall condition. In single-engine aircraft the application of high power will also produce a very strong yawing and rolling effect which, if not combated with sufficient rudder application, can, in some aircraft, particularly with an aft centre of gravity, precipitate a spin entry.

Rudder

This will be needed to combat any yaw at the stall and also that produced by the application of power. On the other hand if it is used too vigorously in a stall situation it could precipitate an entry into the spin, particularly if the control column is moved forward too slowly or insufficiently.

Ailerons

A certification requirement of modern aircraft is that the ailerons must remain effective at the stall. But these certification methods only provide for assessment of stalling characteristics when entered without power. Should the aircraft be using power at the time of entry to the stall and more power is applied during the recovery before the control column is moved forward, the various lateral stability design features, e.g. washout, etc., will nullify aileron effectiveness. This will also apply if a flap is down at the time. Thus, the use of aileron may worsen the situation or at best be of no assistance during the initial stage of the recovery.

In considering this inadvertent stall occurring at low height, we can appreciate that whereas there will be no instinctive opposition on the part of a pilot to use power, rudder and ailerons, there will be a strong resistance to pushing forward on the control column when the aircraft is already moving down to meet the ground with its attendant obstructions. However, if this latter action is not taken the rest of the actions involved in the stall recovery will be valueless.

From these comments stems the real answer to any stall situations, an answer which is encompassed in a simple four-letter word: PUSH. Get this action into effect and, within reason, the rest can be applied without risk a second or so later. This might seem an oversimplification, but it is nevertheless one which stresses the need for the main emphasis to be placed on reducing the angle of attack, an action which clearly was not taken in time in the case of stalling accidents which occurred in the past.

Reducing the risk of a stall/spin accident

There are many causes of this type of accident, and they encompass such items as inadequate preflight preparation, mismanagement of the fuel and other aircraft systems, improper use of the powerplant and aircraft controls (including flaps), and lack of flying skills or good situational judgements.

Lack of good judgement is a fairly common cause and is usually brought about by the pilot not staying ahead of a situation and thus allowing circumstances to develop which are beyond his ability or that of the aircraft to cope with. Examples of this are attempts to take off from fields with inadequate length or pressing on with an approach and landing when all the circumstances are pointing to the need to take 'go-around' action.

The need to stay ahead of the aircraft and maintain a high state of awareness when operating at lower altitudes is of paramount importance if you wish to avoid finishing up as a stall/spin statistic. It will also be of great benefit to hone up one's skills in stall recoveries in the approach configuration from time to time and whilst at a safe altitude.

Too Low
Too Slow

CAUSE:
The probable cause of the accident was that the pilot, who was relatively inexperienced and not qualified to engage in low level operations, allowed the aircraft to stall at a height too low for recovery to be effected.

12
The aircraft may be fit to fly, but what about you?

Pilots who are physically unfit or suffering mental stress are those least likely to use good judgement and arrive at correct decisions. Therefore the last chapter in this book contains some short summaries of various aspects which affect our physical well-being.

Today, no pilot can be considered fully competent unless he has an adequate knowledge of many technical subjects. Some of this knowledge is far removed from the physical act of moving the controls, but in order to ensure that he is capable of correct physical movement at all times during flight he should know and understand the following aeromedical aspects concerning flight.

Just as an aircraft is required to undergo regular checks and maintenance, pilots are also required to undergo regular medical examinations to ensure their fitness to fly. The physical standards they are required to meet are minimum standards. Pilots do not have to be supermen in order to fly. Many defects can be compensated for, as, for example, wearing glasses for visual defects. Pilots may be required to demonstrate by a medical flight test that they can compensate for any other defects of potential significance to flight safety.

While piloting an aircraft, an individual should be free of conditions which are harmful to alertness, ability to make correct decisions, and rapid reaction times. It should be recalled that humans are essentially earthbound creatures. However, if we are aware of certain aeromedical factors, and pay attention to them, we can leave the earth and fly safely. What follows will not be a comprehensive lesson in aeromedical knowledge. It will, however, point out the more important factors which all pilots should be familiar with prior to flying.

Fatigue

Fatigue generally slows reaction time and causes foolish errors due to inattention. The most common cause of fatigue is insufficient rest and loss of sleep; however, others such as the pressures of business, financial worries and

family problems, can be important contributing factors. If your fatigue is marked prior to a given flight, don't fly. To prevent fatigue effects during long flights, keep active in making ground checks, radio-navigation position checks, and remaining mentally stimulated.

It is not always appreciated that fatigue is associated with what is known in aeromedical circles as an 'arousal state'. A pilot's performance is directly related to this state and Fig. 12.1 illustrates the interrelationship between performance and arousal.

In effect this shows that an optimum arousal factor produces an optimum performance, but if the pilot is under-aroused, as in a fatigue situation, or over-aroused, such as when hyperventilation occurs or business pressures are high, his performance deteriorates.

Finally, one very common factor in producing insidious fatigue is the posture of the body, particularly during long flights. Pilots should always take reasonable care to ensure that the seat is adjusted correctly to suit their physical comfort, and the time to do this is whilst on the ground prior to flight.

Hypoxia

Today, in general aviation and within easy reach of all private pilots, there are aircraft which fly above 10,000 feet. Flying at such altitudes can induce a condition known as hypoxia.

Hypoxia in simple terms is a lack of sufficient oxygen to keep the brain and other body tissues functioning properly. Wide individual variation

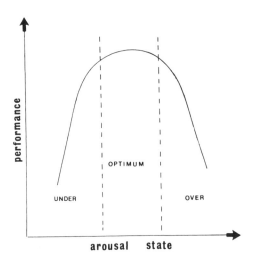

Fig. 12.1

occurs with respect to susceptibility to hypoxia. In addition to the progressively decreasing amount of oxygen at higher altitudes, anything interfering with the blood's ability to carry oxygen can contribute to hypoxia (anaemia, carbon monoxide, and certain drugs). Also, alcohol and various drugs decrease the brain's tolerance to hypoxia. Another quite common factor is temperature; if you feel cold your hypoxia altitude will be lower.

Your body has no built-in alarm system to let you know when you are not getting enough oxygen. It is impossible to predict when or where hypoxia will occur during a given flight, or how it will manifest itself. A major early symptom of hypoxia is an increased sense of well-being (referred to as euphoria). This develops into slow reactions, impaired thinking ability, unusual fatigue, and dull headache.

The symptoms are slow but progressive, insidious in onset, and are most marked at altitudes starting above ten thousand feet. Therefore if you observe the general rule of not flying above ten thousand feet without supplemental oxygen, you should not get into trouble. Night vision, however, can be impaired starting at altitudes lower than ten thousand feet. Heavy smokers may also experience early symptoms of hypoxia at altitudes lower than those for non-smokers. For example, if you are smoking more than 20 cigarettes a day and fly at 10,000 feet without supplementary oxygen the effect would be equivalent to a non-smoker being at 13,000 feet.

Hyperventilation

Hyperventilation, or over-breathing, is a disturbance of respiration that may occur in individuals as a result of emotional tension or anxiety. Under conditions of emotional stress, fright or pain, breathing rate and depth may increase, causing increased lung ventilation although the carbon dioxide output of the body cells does not increase. As a result, carbon dioxide is 'washed out' of the blood. The most common symptoms of hyperventilation are: dizziness, hot and cold sensations, tingling of the hands, legs and feet, nausea, sleepiness and finally unconsciousness. Instrument flying, due to physical concentration and tension, can commonly induce hyperventilation.

Should symptoms occur which cannot definitely be identified as either hypoxia or hyperventilation, the following steps should be taken:

(1) Check your oxygen equipment, put the regulator to 100% oxygen and keep a continuous check on the flow.
(2) If the condition was hypoxia then after three or four deep breaths of oxygen, the symptoms should improve markedly (recovery from hypoxia is rapid).
(3) If the symptoms persist, consciously slow your breathing rate until symptoms clear and then resume normal breathing rate. Breathing can be slowed by talking loudly.

Alcohol

Do not fly while under the influence of alcohol. An excellent rule is to allow twenty-four hours between the last drink and take-off time. This may not be possible but you would be well advised in any circumstances to allow eight hours from 'bottle to throttle'. Even small amounts of alcohol in the system can adversely affect your judgement and decision-making ability.

Remember that your body metabolises alcohol at a fixed rate, and no amount of coffee or medication will alter this rate.

Do not fly with a hangover, or with hangover symptoms suppressed by aspirin or other medication.

Drugs

Self-medication or taking medicine in any form when you are flying can be extremely hazardous. Even simple home or over-the-counter remedies and drugs such as aspirin, antihistamines, or cold capsules, may seriously impair the judgement and co-ordination needed while flying.

If in doubt, seek the advice of your authorised Aviation Medical Examiner. It should also be remembered that the condition for which the drug is required may of itself be hazardous to flying, even when the symptoms are suppressed by the drug.

Certain specific drugs which have been associated with aircraft accidents in the recent past are: antihistamines (widely prescribed for hay fever and other allergies); tranquillisers (prescribed for nervous conditions, hypertension, and other conditions); reducing drugs (amphetamines and other appetite-suppressing drugs) can produce sensations of well-being which have an adverse effect on judgement; barbiturates, sleeping pills, nerve tonics and pills (prescribed for digestive and other disorders) will produce a marked suppression of mental alertness.

Disorientation

The word itself is hard to define. To earthbound individuals it usually means dizziness or swimming of the head. To a pilot it means, in simple terms, that he doesn't know which way is up. So disorientation during flight can have fatal consequences.

On the ground we know which way is up through the combined use of three senses:

(1) Vision – we can see where we are in relation to fixed objects.
(2) Pressure – gravitational pull on muscles and joints tell us which way is down.
(3) Special parts in our inner ear – these tell us which way is down by gravitation pull. It should be noted that accelerations of the body are detected by the fluid in the semi-circular canals of the inner ear, and this tells us when we change position. However, in the absence of visual reference, such as flying into a cloud or overcast, the accelerations can be confusing, especially since their forces can be misinterpreted as gravitational pulls on the muscles and inner ear. The result is often disorientation and vertigo (or dizziness).

All pilots should get an instructor to carry out manoeuvres which will produce disorientation. Once this has been experienced in a dual environment, disorientation which has not been anticipated can be overcome. Closing the eyes for a second or two may help, as will watching the flight instruments, believing them, and controlling the aircraft in accordance with the information they present.

Pilots are susceptible to experiencing disorientation at night, and in any flight condition when outside visibility is reduced to the point where the horizon is obscured. An additional hazard is flicker vertigo, which can lead to confusion and possible disorientation. Light, flickering at certain frequencies, from four to twenty times per second, can produce unpleasant and dangerous reactions in susceptible persons, due to being in phase with the brain's electrical rhythm. These reactions may include nausea, dizziness, mental confusion, interference with concentration, or lead to unconsciousness or reactions similar to epileptic fits. In a single-engine propeller aircraft, heading into the sun, the propeller may cut the sun to give this flashing effect, particularly during landings when the engine is throttled back. These undesirable effects may be avoided by not staring directly through the propeller for more than a moment, and by making frequent but small changes in rpm. The flickering light traversing helicopter blades has been known to cause this difficulty, as has the bounce-back from rotating beacons, and strobe lights on aircraft which have penetrated clouds.

Carbon monoxide

Carbon monoxide is a colourless, odourless and tasteless product of an internal combustion engine and is always present in exhaust fumes. Even minute quantities of carbon monoxide, breathed over a long period of time, may lead to dire consequences.

For biochemical reasons, carbon monoxide has a greater ability to combine with the haemoglobin of the blood than oxygen. Furthermore, once carbon monoxide is absorbed in the blood, it sticks 'like glue' to the haemoglobin and actually prevents the oxygen from attaching to the haemoglobin. The absorption is cumulative and can build up during consecutive flights when causal factors are present; in other words, the normal antidote is time spent in a non carbon monoxide environment.

Most cabin heaters in light aircraft work by air flowing over the manifold. So if you have to use the heater, be wary if you smell exhaust fumes. The onset of symptoms is insidious, with 'blurred thinking', a possible feeling of uneasiness, and subsequent dizziness. Later on, headache occurs. Immediately shut off the heater, open the air ventilators, descend to lower altitudes, and land at the nearest airfield. After this you would be wise to consult an Aviation Medical Examiner, and remember carbon monoxide may take several days to fully clear from the body.

Vision

On the ground, reduced or impaired vision can sometimes be dangerous, depending on where you are and what you are doing. In flying it is always dangerous.

On the ground or in the air, a number of factors such as hypoxia, carbon monoxide, alcohol, drugs, fatigue, or even bright sunlight, can affect your vision. In the air these effects are critical.

Some good specific rules are: ensure windscreens are clean; make use of sunglasses on bright days to avoid eye fatigue; during night flights mask bright torches to avoid destroying any dark adaptation; remember that drugs, alcohol, heavy smoking, and the other factors mentioned above, have early effects on visual acuity. Both distance and depth perception are a function of vision. If you are required to wear glasses, do so and make sure thay are the correct ones.

Middle ear discomfort or pain

Certain persons (whether pilots or passengers) have difficulty balancing the air loads on the ear drum while descending. This is particularly troublesome if a head cold or throat inflammation keeps the Eustachian tube (see Fig. 12.2) from opening properly. If this trouble occurs during descent, try swallowing, yawning, or, as a last resort, holding the nose and mouth shut and gently but forcibly exhaling. If no relief occurs, climb back up a few thousand feet to relieve the pressure on the outer drum. Then descend again, using these measures. A more gradual descent may be tried, and it may be necessary to go through several climbs and descents to 'stair step' down. If a nasal inhaler is available it may afford relief. If trouble persists several hours after landing, consult your authorised Aviation Medical Examiner.

THE AIRCRAFT MAY BE FIT TO FLY, BUT WHAT ABOUT YOU?

Recognition of physical and mental fitness

The feeling of pain is inherent in human beings and this feeling is given to us to act as a warning. If a pain warning is suppressed or ignored we could do untold damage to our physical condition.

If you consider the symptoms by which most of the preceding medical facts and physical malfunctions are initially recognised, you will understand that they are usually in the form of being 'one or two degrees under', and above all, remember that the pilot himself has the sole responsibility for determining his reliability prior to entering the cockpit for flight.

> If you attempt to fly when you are already just 'one degree under' you will not be equipped to recognise and combat hypoxia, hyperventilation, disorientation and similar conditions.

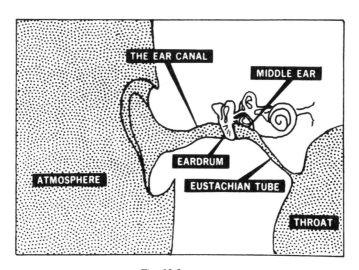

Fig. 12.2

In conclusion

In this book a number of points have been made which, if observed and applied, will be of benefit to the aim of reducing the risks in your flying activities. It is nevertheless acknowledged that there are many other aspects and items of advice culled from experienced pilots which have not been touched upon. This comment applies particularly to IFR operations, but if they had been included, this publication would have assumed encyclopaedic proportions and as such would also have been indigestible.

If some of the points made have provoked thought, and been of help to you, then the objective in writing this book will have been accomplished and I wish you safer flying in the future. However, just before your next take-off, remember to double-check that your seat is securely locked and that your harness and those of your passengers are correctly adjusted and securely fastened. Bear in mind that the lower torso straps should be tightened first and the upper torso straps last. There will be no time to do this if your engine fails immediately after take-off. A recently published NTSB safety report reveals that in survivable accidents, the most common serious injuries are to the head or upper torso; it also notes that 20% of the fatally injured and 88% of the seriously injured occupants had head and upper body injuries which could have been significantly reduced or prevented if only they had been wearing lap and upper torso straps properly!

> **Safe flying makes for greater satisfaction and enjoyment, both of which are secured by knowing and abiding by your own limitations and those of the aircraft you fly.**

Bibliography

UK Civil Aviation Publications
Accidents to Aircraft on the British Register
AIB Bulletins
General Aviation Safety Information Leaflets
Paper 85015: *Human Factors in Accidents*
Paper 85018: *Analysis of Bird Strikes to UK Registered Aircraft*
CAP 507: Aerodrome Operating Minima for Private Pilots
Aeronautical Information Circulars
Paper 85009: *Accidents to UK Aircraft and to Large Jet and Turbo Prop Transport Aircraft World Wide*
CAP 504: *UK Airlines – Annual Operating Traffic and Financial Statistics Safety Performance of UK Airline Operators*
Paper 82013: *Analysis of Occurrence Data*

US National Transportation Safety Board:
Annual Reports to Congress
Annual Reviews of General Aviation Accidents
Special Study of Fatal Weather Involved General Aviation Accidents
Special Study of US General Aviation Accidents Involving Fuel Starvation
Special Study Report on Approach and Landing Accidents
Special Study of Accidents Involving Engine Failure/Malfunction to US General Aviation Aircraft
Special Study of Midair Collisions in US Civil Aviation
Special Study of General Aviation Stall/Spin Accidents
Safety Report – General Aviation Crashworthiness Project

Federal Aviation Administration:
Accident Prevention Information Leaflets
General Aviation News Publications
Aviation Advisory Circulars

Australian Bureau of Air Safety Investigation:
Annual Surveys of Accidents to Australian Civil Aircraft
Aviation Safety Digests

International Civil Aviation Organisation:
ICAO Statistical Year Books
Civil Aviation Statistics of the World
ICAO Bulletins
Aircraft Accident Digests
Traffic by Flight Stage
Manual on Aerial Work

Aircraft Owners and Pilots Association (AOPA):
AOPA Flight Safety Foundation – *Special Study on General Aviation Safety*
Flight Instructors' Safety Reports AOPA (US)
AOPA UK Instructor Bulletins

UK General Aviation Safety Committee:
Flight Safety Bulletins
Special Reports

Index

ability – overload, 26
accident briefs – weather, 135
accuracy landings, 35
adverse weather, 133
aerodrome approach checks, 85
aeromedical factors, 187
aircraft manual, 36
aircraft performance, 36
airframe icing, 129
airspace system, 12, 32
altimeter settings, 86
altitude – temperature, 45
ammeter charge, 65
angle of attack, 66, 73
arousal state, 188

backside – power curve, 97
base leg – procedures, 90
battery charge, 65
birdstrike
 encounter, 79
 statistics, 72
brake pressure checks, 87
breakaway thrust, 72

calculations
 take off, 47
 landing, 51
carburettor heater, 160
 icing, 64, 104, 153
carelessness, 18
centre of gravity, 36, 183
checklists, 61
cockpit workload, 144
cognition, 13, 20
collision with objects, 25
complacency, 18, 103
compulsion, 18, 43, 55
continued flight below minima, 123
crosswind factors accident briefs, 70
cumulonimbus, 139

de-icing equipment, 39, 129
density altitude, 39, 129
design induced error, 26

desolate areas, 34
detonation, 161
disorientation, 128, 191
distraction – pilot induced, 145
downdraughts, 95

engine malfunction, 25
engine failure
 accident briefs, 118, 163
 remedies, 172
evapouration process, 157
excess thrust horse power, 95
exhaust muffs, 160

flaps, 55, 65
flicker vertigo, 191
flight plan, 35
flight planning, 30
flight phases, 20, 22
float period, 92
fuel
 consumption, 107
 contamination, 115, 150
 drains/strainers, 114
 exhaustion, 30, 105
 flow gauge, 106
 grade, 113
 mismanagement, 30, 103
 mixture, 46
 pumps, 106
 related accidents, 104
 starvation, 105
 tank selection, 85, 105
 quantity gauge, 110
frost, 37

gear
 collapsed, 25
 landing, 16
 up incidents, 25
go-around, 18, 92

habits, 27, 56
halfway rule, 54
hard landings, 25

INDEX

heading indicator, 65, 86
'hot and high', 91
humidity, 46

ice, 37, 41
idle power, 72
impact icing, 157
inadequate distance, 25
inadvertent stalls, 185
induction air, 157
induction system, 157
inlet manifold, 157
insidious fatigue, 188
instrument flying, 39

jet blast, 72
judgement, 13, 18, 27, 43

key-type ignition, 174
knowledge, 20
Koch chart, 50

landing accident briefs, 57
   gear, 66
lift off speed, 52
line squalls, 129
long grass, 67
lookout, 88
loss of control, 25

magneto checks, 63
maximum rated power, 39
memory, 26
mental preparedness, 145
meteorological charges, 130
midair collisions, 85, 88
minimum control speed, 39
mistakes, 27
mixture control, 104
motor skill, 15, 18, 183
mountainous areas, 34
multiple tasks, 27

navigation log, 35
no-go situations, 39, 67, 128
Normal Category, 182
nose over, 25

overboosting, 39
over runs, 25, 91

performance factors, 44
pilot privileges, 128
pilot distractions, 143

pilot heater, 130
power curve, 97
power plant failure, 117
precision approach, 91
pre-flight inspections, 148
pre take-off checks, 65
prevention – accident, 147
propellor failure, 62
   slipstream, 71

radio navigation, 39
rate of climb, 48, 95
reversed command, 97
risk factors, 21, 22
route planning, 34
runway
   conditions, 38
   gradient, 38

safety altitude, 126, 139
Safety Data Analysis Unit, 148
self discipline, 27
short field
   take-off, 43, 53
   landing, 43
simulated emergencies, 57
slipstream, 71
soft ground, 67
stall/spin accidents, 143, 180
stall/spin awareness and avoidance, 180
stall recovery, 183
stalling speed, 95
surface conditions, 46

take-off
   calculations, 47
   accident briefs, 57
temperature/altitude, 45
throttle butterfly, 157
throttle icing, 158
thrust/weight ratio, 45
thunderstorms, 129, 140
touchdown zone, 74

undershoot, 91
use of
   charts, 34
   flap, 65, 93
Utility Category, 182

valve – throttle, 160
venting systems – fuel, 116
venturi, 157
visibility, 127

INDEX

Vmc, 39
VMC, 39
vortices, 73
vortex strength, 73
Vso, 52, 68
Vsi, 52

wake turbulence, 71
wake vortex encounters, 78
water vapour, 46

weather
    deterioration, 32, 123
    fatal accidents, 133
    forecasts, 35, 139
    related accidents, 25
    reports, 35
weight and balance, 36
wind
    effects, 38, 45
    gradient, 175
    gusts, 25, 37
    shear, 142